ROOFING

THE BEST OF
Fine Homebuilding

ROOFING

THE BEST OF
Fine Homebuilding

The Taunton Press

Cover photo: Scott McBride

**Back-cover photos: Bethlehem Steel (top),
Cedar Shake and Shingle Bureau (center),
Jefferson Kolle (bottom)**

Taunton
BOOKS & VIDEOS

for fellow enthusiasts

First printing: 1996
Printed in the United States of America

A Fine Homebuilding Book

Fine Homebuilding® is a trademark of The Taunton Press, Inc.,
registered in the U.S. Patent and Trademark Office.

The Taunton Press, Inc.
63 South Main Street
P.O. Box 5506
Newtown, Connecticut 06470-5506

Library of Congress Cataloging-in-Publication Data

Roofing : the best of Fine homebuilding.
   p.   cm.
  Includes index.
  ISBN 1-56158-141-0
  1. Roofing.   I. Taunton Press.   II. Fine homebuilding.
TH2431.R54   1996
  695 — dc20   96-5432
                                                CIP

Contents

INTRODUCTION

A FRIEND OF MINE says that when you fly over Oregon, it looks as though everyone has a swimming pool. But what you're really seeing are houses with huge blue tarps thrown over them to keep their roofs from leaking. Now, I don't think my friend meant to disparage Oregon—in fact, he's moving there himself. You can probably find the blue-tarp solution at work in any state.

But if you'd rather not have a roof that flaps in the wind like a bed sheet on a clothesline, if you'd prefer a more permanent, less colorful solution to the problem of keeping rain out of your house, this book has the answers. Collected here are 19 articles from past issues of *Fine Homebuilding* magazine, covering every aspect of roofing from choosing a roof in the first place to installing whatever you choose, be it asphalt, metal, concrete, tile, cedar or even thatch.

—Kevin Ireton, editor

Choosing Roofing

Up in the air about using asphalt, wood, slate, tile or metal to keep out the rain? There are lots of choices, and they all have advantages

by Jefferson Kolle

A lot of roofing materials try to look like something else. When they first came out, asphalt shingles were touted as looking like slate. Today, there are metal roofs that are supposed to look like tile, and there are fiber-cement roofs that are supposed to look like wood shingles.

Whether you're looking for a roof that looks like something else or a roof that looks like what it is—wood roofs really do look like wood—there are a lot of materials on the market, and they all have their benefits. Material costs and installation costs of some roofing materials are higher than others, but the payback is in their longevity or in their aesthetic appeal. What follows is a survey of the most common roofing materials available for steep-roof residential construction (anything greater than a 3-in-12 pitch).

The standard unit of measurement for roofing materials is the square. A square is 100 square feet of roofing. Manufacturers refer to their products on a per-square basis—cost per square, weight per square, etc. This article will use the same nomenclature.

Asphalt roofing is inexpensive and can be installed quickly—One story has it that the three-tab asphalt strip shingle, with its two grooves dividing the exposed face of the shingle into three sections, was invented by Fred Overbury in 1915 when he pulled a cardboard divider out of an egg crate and was struck with a brilliant idea. Before Overbury's invention asphalt shingles were made as individual pieces and were installed one at a time. The strip shingle revolutionized the asphalt-roofing industry.

Today, asphalt shingles cover more residential roofs than any other material. Every year, 100 million squares of asphalt shingles are installed in the United States. That's more than 358 square miles, about the size of Lake Tahoe.

Asphalt shingles' popularity is due to several factors. They're fire-resistant, and choices of color and textures are numerous. They're relatively inexpensive both to purchase and to install; an experienced roofer can install 10 to 20 squares a day, depending on the intricacies of the roof.

On a sunny summer day, a black asphalt roof can reach a temperature of 150°F. As the temperature rises, asphalt shingles become soft and pliable. A sudden thunderstorm can cause the temperature of that soft, pliable roof to drop to around 60°F. That's called thermal cycling. If the asphalt shingles on the roof are going to continue to shed water, their reaction to thermal cycling has to be minimal. Of course they will shrink when the temperature drops. But they can't curl, and they can't lift off the shingle below.

All asphalt shingles share common construction: A reinforcing mat is impregnated with asphalt. Twenty-five years ago, fiberglass mats were

---

Each year 100 million squares of asphalt shingles are used in the U.S. That's an area about the size of Lake Tahoe.

---

introduced to replace the earlier organic-fiber mats. Although organic-fiber mat shingles still are recommended for areas with extreme winds, early blow-off problems with fiberglass-mat shingles have been eliminated so that, today, fiberglass-mat shingles are the most common type sold.

Filler materials, most commonly ground limestone, help stabilize the shingle's asphalt—technically a liquid—by stiffening it and keeping it from flowing. The fillers' inertness adds to the fire retardancy of the shingles, and they increase resistance to cupping during thermal cycling.

Asphalt degrades in sunlight; it loses its suppleness, dries out and cracks. To combat that problem, asphalt shingles have a surface coating of granulated minerals pressed into the part of the shingle exposed to the sun. Eventually, when the

mineral granules wear off a shingle, through abrasion or erosion, the shingle degrades quickly. The granules are what give a shingle its color. Colors from bright greens to blues, yellows and reds are available as well as blacks, whites and a variety of subdued earth tones.

Architectural shingles are thicker than three-tabs—About the time fiberglass reinforcing mats were introduced, manufacturers came out with what are known as architectural or laminated shingles. Unlike three-tab shingles with cutout grooves, architectural shingles typically are solid across their length. Multiple, overlapping layers are laminated to form a heavier shingle with a more textured appearance.

Some architectural-shingle manufacturers use different colored mineral granules on the multiple layers to form the illusion of the cast shadowlines (top left photo, facing page) one might see on a wood-shake or slate roof. From the sidewalk, at dusk, in the fog, one might think a cedar-colored architectural-grade asphalt-shingle roof was wood. Other than that, they aren't convincing. The slate imitators also are unconvincing.

Most three-tab shingles weigh around 240 lb. per square, but some architectural shingles can weigh as much as 100 lb. more per square. Three-tab shingles come with a 15-year to 20-year warranty, but, because there is more material in the architectural shingles, they come with a longer warranty, typically 30 to 40 years (bottom photo, facing page). Architectural shingles are sold at a premium. In Newtown, Connecticut, for instance, three-tab shingles sell for $24 per square; architectural shingles cost $52 per square.

There's a common thought that a heavier shingle is a better shingle. But according to a paper sponsored by the National Roofing Contractors Association, "shingle testing and observations from field performance have frequently shown that weight alone is not a sufficient indicator of shingle quality..." Rather, "the quality of the individual components of the composite structure—the reinforcement [mat], the asphalt and the filler—are much better indicators of shingle performance." According to W. Kent Blanchard, one of the authors of the report, there is no easy way for a consumer to get an indication of the quality

Architectural asphalt shingles try to look like wood. Different colored surface granules and overlapping layers on thick architectural shingles attempt to mimic the textures of a wood-shingle roof. Photo courtesy of Certainteed Corporation.

The look of real wood shingles. The natural beauty of a wood roof palliates its high cost. These pressure-treated red-cedar shingles are warranted for 30 years against fungal decay. Photo courtesy of The Cedar Shake and Shingle Bureau.

Architectural shingles cost more but last longer. Architectural asphalt shingles cost twice as much as three-tab shingles. The advantages are a longer warranty and the expectation of a longer life. Photo courtesy of Certainteed Corporation.

Clean keyways add life to wood roofs. Power washing, broom sweeping or rinsing with a hose will clean rot-causing debris from between shingles. Periodic treatments with wood preservative will help extend the roof's life. Photo courtesy of The Cedar Shake and Shingle Bureau.

Metal that tries to look like tile. Painted-metal panels weigh only 125 lbs. per square. Their eaves-to-ridge length makes for fast installation. Photo courtesy of Met-Tile Inc.

The look of a real tile roof. The color of clay-barrel tiles won't fade in the sun. Photo courtesy of M.C.A. Inc.

Aluminum shingles are light. Coated aluminum shingles weigh only 50 lbs. per square. Touted as looking like wood, their textured appearance is unique. Photo courtesy of Alcoa Building Products.

of a shingle's components, but he said to make sure any fiberglass-mat shingle you purchase has passed ASTM standard D6432, a measurement of tear strength. According to an article in the September 1992 issue of *Roofer Magazine*, "High tear strength is a good indicator of shingle toughness and resistance to cracking."

Wood roofing is beautiful—Regardless of manufacturers' attempts to simulate the appearance of wood shingles, nothing really looks like a wood roof except the real thing (top right photo, p. 9). Although wood roofs are expensive to purchase and to install, many people think their aesthetic value outweighs their high cost.

Wood roofs are composed of either shingles or shakes, and, although both are wedge-shaped in section and often are confused for one another, there are basic differences that set them apart. Put simply, a shingle's tapered shape is attained by sawing, and a shake's is split, or rived, from logs. Shakes are thicker at the butt, or bottom, than shingles, and the striations that result from splitting along the length of the grain give shakes a more rustic and textured appearance than sawn shingles.

Wood roofs are not only expensive to purchase—a square of top-quality shakes can cost more than $150—they also are labor intensive to install. Because wood shakes and shingles are nailed onto a roof one at a time, it takes a while to cover a roof with wood. A fast shingler on a section of roof without any time-consuming obstructions to roof around can put on 2 or 3 squares per eight-hour day.

Preparation of the deck before roofing can begin also takes a while. Good practice dictates laying shingles or shakes over skip sheathing, solid boards nailed with spaces between the rows, rather than over plywood or a similar solid decking. Skip sheathing is preferable for a wood roof because the air spaces between boards allow the wood roofing to dry from both sides, thus contributing to the roofing's longevity.

Maintaining a wood roof—Except for certain metal roofs that can be repainted, wood is one of the few roofing materials that can be maintained. And, in fact, it is recommended that wood-roof maintenance will protect an expensive investment. Brian Buchanan is a wood technologist in Lufkin, Texas. He has written a treatise called *Evaluating Various Preservative Treatments and Treating Methods for Western Red Cedar Shingles.* Buchanan told me that the most important practice toward promoting a wood roof's longevity is to keep the keyways between shingles clear of debris (top photo, left). When the keyways get clogged—with leaves, conifer needles, dead birds, whatever—fungus starts its eventual course toward rot. Power washing, sweeping with a broom and even simple washing with a garden hose cleans out keyways. Keep in mind that a wet wood roof is a slippery wood roof.

Liquid preservatives for wood roofs vary in their effectiveness. The best contain copper naphthenate. While some might consider the green color of products containing this compound to be unsightly, there are pigmented versions that simulate cedar's reddish-brown color. A quick read of a can's label can tell you if the product contains copper naphthenate. A free copy of Buchanan's pamphlet is available from the Texas Forest Service (409-639-8180).

Western red cedar covers more roofs than any other species. Alaskan yellow cedar shingles are available as are ones made of Eastern white cedar. And, recently, the Southern Pine Marketing Council (504-443-4464) has begun touting pressure-treated Southern pine shakes.

Wood roofs and fire—Red-cedar shingles that have been impregnated with a fire retardant are given a Class B fire rating (Class A is the most fire resistant). A Class A roof can be had using wood, according to The Cedar Shake and Shingle Bureau, but you have to install the fire-resistant shingle over a sandwich of two layers of sheathing with a layer of ½-in. gypsum board between.

Fire retardants work well; the wood won't burn after it is treated. The problem comes from the treated wood's exposure to weather. Rain soaks the wood roof, and, when the sun comes out and dries the wood, the retardants are drawn to the surface. Subsequent rains wash the retardant off the wood. The process is repeated over the years, and, when all the retardant leaches out of the wood, your roof is covered with kindling.

But The Cedar Shake and Shingle Bureau has another view of the situation. Don Meucci, a spokesman at the bureau, said that tests done of fire-treated wood shingles taken from a roof 16 years after installation passed the same stringent tests that new shingles must undergo today.

But communities across the country from Los Angeles, California, to Newcastle, New Hampshire, have banned wood roofs, even those treated with fire retardants, because of their flammability. Of course, if a fire starts in your kitchen, no roof is going to keep your house from burning. The problem with wood roofs is twofold: Sparks landing on a wood roof can cause it to burn, and when the wood roof catches fire, the wood can send off flying brands, bits of burning material that leave the roof with the smoke column and then fall to the ground, still glowing hot, ready to start the next fire.

If you're considering a wood roof, talk to The Cedar Shake and Shingle Bureau (206-453-1323), your building department or your local fire chief.

Metal roofs aren't just for barns anymore— My first memory of metal roofing is not a pleasant one. My brothers and I were on my grandfather's farm, and we discovered that the big barn was full of bats. One of us had the brilliant idea to rid the barn of the flying rodents with bow and arrows. We didn't hit any bats, but we did puncture the roof three or four times. My grandfather saw the arrows sticking though the galvanized sheets of roofing. He was not pleased.

Metal roofing is no longer relegated just to farm buildings, and it is available in styles and colors other than the rusting galvanized tin seen everywhere from Walker Evans' early photographs of sharecroppers' houses in Tennessee to my grandfather's barn in southern New Jersey. Improvements in both metal-coating processes

COMPARISON OF ROOFING MATERIALS

	Asphalt	Wood	Metal	Tile	Slate
Cost/square	$25-$56	$150-$200	$35-$250	$120-$1,000	$350-$700
Installation* cost/square	$65-$125	$130-$160	$35-$400	$100-$300	$250-$450
Approx.** life span/yrs.	15-20	10-40	15-40+	20+	30-100
Weight in lb.	225-385	300-400	50-270	375-1,100	500-1,000
Fire rating	A	B***	A	A	A

*Installation costs vary enormously due to many factors such as local labor rates, time of year, complexity of a house's roof geometry, height of a roof from the ground and complexity of a roofing material's profile.

**Roofing materials' life spans are courtesy of the American Society of Home Inspectors.

***Wood shingles and shakes treated with a fire retardant are given a Class B fire rating. Untreated shakes and shingles have no fire rating. A Class A roof is attainable with wood roofing, but a special installation procedure involving a sheathing sandwich made of plywood and gypsum board is necessary.

Standing seam can look great on a home. The vertical ribs of a standing-seam roof give the metal panels rigidity. The slick surfaces shed water and snow quickly. Because of this, gutters should be sized accordingly or, if possible, eliminated. Photo courtesy of Bethlehem Steel.

Slate roofs can last indefinitely. The fasteners and flashing will wear out before the roof slates will. Because of slate's long life, there is a thriving market in used slate. Photo by Terry Smiley.

and in waterproof fasteners have made metal roofing suitable for residential use. According to Sig Hall, an estimator for Bryant Universal, the largest roofing contractor in the United States, metal roofs are being put on homes in a number that is increasing faster than any other material.

Metal roofs have been around for a long time. First fashioned on site by hand and later manufactured in sheets installed in long sections, metal roofing is available in patterns other than standing seams or sine-wave corrugations. Some manufacturers make panels that are supposed to simulate clay tiles (center left photo, p. 10).

A variety of metals is used for roofing—everything from copper to stainless steel to aluminum to alloys and coatings and compounds of each. Painted finishes are available in a range of colors as wide as what is offered on today's cars. Most manufacturers will blend custom colors if you're really picky and can't find something in their stock palette. Because of its thinness, metal roofing is the lightest of all roofing material. There are coated aluminum shingles that weigh only 50 lbs. per square (bottom photo, p. 10). Because metal roofs are so light, the metal-roofing industry touts metal as an excellent choice for reroofing. Depending on the application—some materials might require the addition of furring strips or standoffs—a metal roof can go over an existing asphalt or wood roof.

The metal-roofing folks also harp on their product's ecological benefits. Because no tearoff is required, no shingles, asphalt or wood, go into landfills. Another benefit to the environment is that metal roofing is the only completely recyclable product. When a metal roof does wear

out, and most metal roof are expected to last 50 years, it can be recycled into new roofing. And there's a good chance that the metal roof you put on your house used to be a beer can.

Some manufacturers make metal shingles, but most metal roofing comes in panels that go on quickly. Because of the long eaves-to-ridge panels that are available—some manufacturers have shipped panels 80-ft. long—a lot of square footage gets covered at a time. Depending on the complexity of a roof's geometry and the complexity of the profile of the metal roofing being used, an experienced installer can lay down between 3 and 30 squares in a day.

The cost of a metal roof depends on the kind of metal used, the thickness of the material and the finish. According to Rob Haddock, the director of the Metal Roof Advisory Group, the least-expensive metal roof—galvanized tin sheets—can be purchased and installed for as little as $60 per square, while handcrafted copper on a complex roof could cost as much as $1,500 per square. In an article in *Contractors Roofing and Building Insulation Guide*, Haddock wrote that "metal roofing is the lowest cost, highest cost and everything-in-between roofing material."

Standing-seam roofs are the most popular profile (photo, p. 11); they have a vertical seam that stands proud of a flat panel. Panels are joined together at the edge seams either by a crimping of the seams or by a cap that covers them. A standing-seam panel can be formed either at the factory or on site by a portable roll-forming machine that bends a coil of metal.

If you live in snow country, keep in mind that metal roofs, especially eaves-to-ridge panels,

shed snow quickly, kind of like sledding in reverse. Snow cover on a metal roof will have a tendency to let go: That is, it will slide off the roof all at once, like a miniature avalanche. It might be a good idea to keep your foundation plantings 3 ft. or 4 ft. away from the house if you don't want to lose that prize pyracantha to snow damage. Slick metal roofs also shed rain faster than some other, more textured roofing materials. Because of this, gutters need to be sized accordingly or, if possible, eliminated.

Tile roofing—Most people think of tile roofs as indigenous to Florida and the Southwestern part of the United States and as appropriate only to Spanish-style or Mediterranean-style houses. But tile roofs are popular in Europe. And in Japan, tile's popularity rivals that of asphalt in this country. In Japan there are 500 companies making roofing tiles. In this country there are five.

Tile roofs have a textured look. Flat tiles are available, but the soft undulations of a barrel-tile roof or the crested-wave appearance of a Japanese tile roof are markedly more textural than the appearance of most other roofing materials, which aside from the slight deviation of course lines or seams, are planar in appearance.

Whereas most other roofing materials are classified by the dominant material in their composition—wood, asphalt, metal—roof tiles are referred to as such because of the process of their manufacture. Like interior house tiles, roof tiles are made of a soft, plastic material—either clay, concrete or fiber cement—that is molded or extruded and then hardened into a brittle, inert state either by heat or by chemical reaction.

Because of its inertness, tile won't burn. You'd have the same difficulty getting a roof tile to burn as you would if you tried to ignite a concrete block or a clay flowerpot. All roof tiles have a Class A fire rating. After the 1992 fires in southern California, some of the only houses left standing had tile roofs. No house is immune from fire, but sparks on a tile roof won't cause immolation.

The question of weight always arises when people talk about tile roofs. There are glazed-clay tiles that weigh more than 1,100 lb. per square, but keep this in mind: Three layers of asphalt shingles (an original and two reroofs) can weigh about 900 lb. And all new-house roofs are engineered to carry three asphalt roofs. On the other hand, there are lightweight concrete tiles that weigh as little as 375 lb. per square. If you are considering a tile roof and if you have any questions about your house's ability to withstand the weight, talk to an engineer. He can assuage your doubts or, possibly, suggest some roof reinforcement that might not be as expensive as you think.

Are roof tiles expensive? Well, how much does a new car cost? Both questions need qualification before an accurate answer can be given. According to Stu Matthews, owner of Northern Roof Tile Sales in Ontario, Canada, the price of a roof tile is not an accurate indication of its quality. The quality of barrel tile costing $600 per square is similar to tiles with the same profile that cost $120 per square.

If price does not indicate quality, what does affect the price of tile? Matthews said there are several things that affect cost. As in most commodities, a manufacturer's production volume affects the price at which it can offer its goods. A company with automated manufacturing procedures, making 150 million tiles per year, can sell its tiles for a lot less than a manufacturer making ¹⁄₁₀ that number.

What else affects cost? There's a British saying: "You can tell a person's wealth by the size of the tiles on his roof." Contrary to the popular notion that bigger is always better and more expensive, in the case of roof tiles the opposite is true. On a per-square basis, smaller tiles cost more to purchase and to install. The smaller the tile, the more tiles there are per square, and, because tiles are installed one at a time, smaller tiles are labor intensive to put on a roof. The most expensive tile that Matthews sells is handmade, and there are 500 tiles per square. They sell for about $1,000 per square. As a point of comparison, there are terra-cotta roof tiles, available in California where they are made, that have 75 tiles per square and cost about $75 per square.

Because roof tiles are heavy and because there are a limited number of manufacturers in this country, shipping costs to a job site *can* have a huge effect on the cost of a tile roof. For instance, the same $75-per-square tile mentioned above could double in price by the time it reaches a job in Massachusetts.

Clay, concrete or fiber cement?—Unlike interior tiles, which are made, for the most part, of a natural ceramic material such as clay or porcelain, roof tiles can be made of concrete or fiber cement. All materials have distinct advantages.

Clay was the first material used for roof tiles, and, in this country, it is still the most popular (center right photo, p. 10). Terra-cotta tiles are flower-pot color and commonly are used on a roof in their natural color. Terra-cotta can be colored by different methods. Engobe is a process in which a colored wash is put on tiles. When raw tiles are fired in the kiln, they take on the color of the wash. Engobe-fired tiles are limited to muted earth tones. Glazing is another process altogether. After an initial firing, tiles are coated with a glaze and fired again. Bright, primary colors are standard offerings from manufacturers that sell glazed tiles, and a lot of companies make custom colors on request. Also, clay tiles won't fade in the sun like concrete tiles.

A final note on ceramic tiles: Celadon (P. O. Box 309, New Lexington, Ohio 43764-0309; 800-235-7528) is an interlocking ceramic tile that looks like slate (bottom photo), but not in the way that some asphalt shingles are supposed to look like wood. Rather, these ceramic tiles have, to my eye, a genuine slate appearance.

Concrete tiles are less expensive than real clay tiles; some cost as little as $50 per square. They are available in profiles that simulate clay-roof tiles. Imitations of slate, wood shakes and wood shingles are all available in concrete. Concrete tiles are heavy—around 900 lbs. per square—but Westile (8311 W. Carder Court, Littleton, Colo. 80125; 800-433-8453) makes a concrete tile with the paradoxical name FeatherStone that weighs 660 lb. per square.

One is slate, the other is molded clay. The photo at the top is of a real slate roof, and the photo on the bottom is of a Celadon clay-tile roof. Or is it the other way around? Top photo by Terry Smiley; bottom photo courtesy of Certainteed Corporation.

Concrete can be colored, but air pollution fades colored concrete. Matthews recommended using concrete tiles that have the color mixed through the concrete, rather than ones that have a wash of pigment applied to the surface.

Fiber-cement tiles have been around for a long time. Unfortunately, the fiber in fiber-cement tiles used to be asbestos. Manufacturers are vague about what replaced asbestos in fiber-cement tiles, but the new ones contain no asbestos. Fiber-cement roofing tiles are a lot lighter than either concrete or clay tiles. Cembrit (170 Ambassador Drive, Mississauga, Ont. L5T 2H9; 905-564-3110) makes a roof-tile panel that weighs only 350 lb. per square and comes with a 30-year guarantee. Fiber-cement tiles cost between $225 and $275 per square, and they are made to imitate clay tiles, slates, wood shakes and wood shingles.

With the ban on wood roofs in many parts of the country, fiber-cement roof tiles that imitate wood roofs are catching on (instead of catching fire). The Clarke Group (P. O. Box 1094, Sumas, Wash. 98295; 800-347-3373), which touts itself as "the world's largest manufacturer and treater of cedar shakes and shingles," recently has come out with a fiber-cement roofing product that imitates wood shakes and shingles. The name of the product? FireFree.

In areas outside Florida and the Southwest, tiles are beginning to become popular. If a roofer tries to convince you that you don't want a tile roof, ask him how many he's installed. Chances are, his experience is limited. Some tiles are more difficult to install than others, but the hardest roof you'll ever tile will be the first one. Depending

on the tile—the installation of some are more labor-intensive than others—an experienced roofer should be able to put on between 2 and 3 squares per day. Tile manufacturers are excellent at disseminating installation literature.

Slate roofing can last hundreds of years—In London, there's a building called Westminster Hall that was finished at the turn of the 10th century. They put a new roof on the place in the 13th century, and a smart contractor chose slate. The same roof is still on the building.

Roof slates should never wear out (photo, facing page). Given that the material is already a couple of million years old, expecting it to last another hundred years or so on your roof isn't really asking a lot. What do fail are the fasteners that hold the slates to the roof and the flashing at junctures such as valleys and around chimneys.

Because of its longevity, slate is the only roofing material that sometimes can be purchased used. Slates carefully are removed from a roof, the cracked or damaged ones are culled out, and the remaining slate can be reinstalled on another building. Can you imagine putting used asphalt or used wood shakes on a roof?

Slate is expensive. A square of slates can cost between $350-$700. Add to that the expense of installation, between $250 and $450 per square, according to Terry Smiley, a slate roofer in Denver, Colorado, and you've got an expensive roof. But divide the per-square price by 100 years, and it doesn't seem so expensive.

A slate roof might not be as heavy as you think. Depending on the slate's thickness—³⁄₁₆ in. is the industry standard (top photo)—a slate roof can weigh between 650 lb. and 1,000 lb. per square.

Care must be taken when nailing slates on a roof. Sort of the Goldilocks syndrome: not too hard, not too soft. Slates have to be nailed just right. Nailing a slate too tightly will cause it to crack when the slate expands and contracts. And nailing a slate too loosely can cause the slate above to crack. Bill Markcrow at Vermont Structural Slate in Fair Haven, Vermont, favors slate hangers. Hangers have been around for a hundred years, but they had fallen from popularity. Markcrow thinks slate hangers are foolproof. They have a hook on one end and a nail on the other. Using them is easy; you still have to nail the slates around the perimeter of the roof, but, for the rest, you just nail on the hangers and place the slates on the hook. The hangers are made of stainless steel, and you can get them painted black. Markcrow says the hangers are visible hooked under the bottom of the slate, but "from 8m away, they disappear."

It's not likely that you'll be able to go to your local roofing-supply store and take home enough slates to roof your house. Slate will have to be ordered from a quarry that will cut it from the ground and then fabricate the material into roof slates. Expect a month or so between the time you place an order and get delivery. But in the larger time frame of a slate roof's life, what's another month? □

Jefferson Kolle is an associate editor at Fine Homebuilding.

Take it all off. When the second layer of asphalt shingles started leaking, it was time to tear off the roofing down to the old board sheathing. After tacking a tarp along the fascia to protect the house and the grounds, the roofers pulled off the shingle caps, then worked their way down with ripping shovels, straight-claw hammers and pry bars. Triangular brackets are driven between sheathing boards to provide footholds.

Tearing Off Old Roofing

The best reroofing jobs start with a clean roof deck

by Jack LeVert

Twin brothers Richard and Russell Wright have reroofed hundreds of Boston-area houses since they went into business together 25 years ago. I've been up on the scaffolding with them on quite a few. I usually do the specialty carpentry—gutters, fascia, skylights, sheathing repairs—but when it's time to tear off an old roof, everybody pitches in.

Tearing off a roof (photo facing page) is a messy, nasty job. You've got to take steps to protect the house, the grounds and yourself. Here I'll explain how to determine whether a house needs a new roof and what to do if it does.

Inspecting the roof—When checking out a leaky roof, I first look for structural damage to the roof itself. I do it from inside the attic. I look for signs of continuous moisture, such as water stains, patches of dry rot or black fungus. And I probe for rot. With an awl, I poke the underside of the sheathing and the top edges of the rafters. Softness indicates water damage. If the interior damage is limited to one spot, this spot may be the source of a leak, and depending on the condition of the shingles, I might simply choose to patch the leak.

If the rafters are sagging, they were probably undersized when the house was built. I jack up sagging rafters and sister on new rafters. This is done before tearing off the roofing; otherwise, jacking up the rafters could spring loose the newly installed roofing or even the sheathing.

Sagging between rafters indicates a problem with the sheathing: It may be rotted or cracked; it may be undersized; it may have been run across too few rafters to provide strength; or it may not have been staggered across the rafters properly. A second layer of sheathing often will correct this problem.

Extensive fascia and soffit damage indicates that either the drip edge is bad, or the rafter tails have rotted. I pull off a section of damaged fascia to check out the rafters. Although I sometimes remove the sheathing to replace rotted rafters, often I can repair rafter tails without removing the roofing. If the rafters are OK, a defective drip edge is probably causing the water damage, and a new drip edge will have to be installed with the new shingles.

If there's any sign of carpenter ants, I tear off some sheathing, determine the extent of the infestation and get rid of it. The best way to do away with carpenter ants is to remove all wet wood, whether it's infested or not.

What to look for on the roof—Like car batteries, shingles are rated to last a certain length of time. Fortunately, shingles are rated in years, not months. You can buy shingles rated to last anywhere from 15 years to 30 years. But you can't determine whether a roof needs to be replaced simply by comparing the age of the shingles to their projected life span. I've heard of shingles deteriorating in as little as five years, and I've seen roofs that have lasted well over 30 years.

You must go on the roof and look for signs of deterioration (photo above)—shingles that have lifted and curled edges, brittle shingles that crum-

Past their prime. **These deteriorated asphalt shingles have missing corners, and roofing nails show. They also snap and crumble easily, and the black showing through indicates that the protective layer of granules has worn away, leaving the asphalt exposed to the sun's ultraviolet rays.**

ble when lifted. On a cold day, even a new shingle will be brittle and will snap in your hand, but in the old, deteriorated shingle's case, the exposed edges will crack and crumble like stale cake. Here and there a deteriorated roof will look like the worn soles of old shoes, with round holes revealing the shingles underneath. Corners of shingles will be missing, and the heads of roofing nails will show—sure signs of leakage.

In areas that get little or no sun, shingles remain wet for a long time and may be green and slick with moss. Water-damaged shingles will be soft and mushy. If you can leave a thumbprint in a shingle, the shingles are too far gone to keep.

In winter, snow melts slowly in these shaded areas. The constantly trickling water freezes, melts and refreezes, wearing away the protective layer of granules on asphalt shingles. The condition of the granules impregnated in the shingles is the real key to determining whether an asphalt roof is shot. Asphalt shingles consist of a blend of steep asphalt (asphalt that won't soften below a temperature of 130° F) held together with strands of fiberglass and covered with a top layer of ceramic-coated granules. Each of these three components has a function. The asphalt protects the roof; the inorganic fiberglass mat strengthens the shingle; and the granules shield the asphalt from the sun. (The granules are crushed rock that resist ultraviolet light, heat up to 1,200° F and acid.) Rain, snow, changes in temperature—all weather—gradually wear away this protective layer of granules. Once the sunscreen of protective granules has worn off, the sun's ultraviolet rays evaporate the natural oils of the asphalt, causing the shingle to degrade.

Roofing over an old roof—In Massachusetts, where we work, the building code allows reroofing over a single layer of shingles but not a third roof over two. This is a good rule to follow, whether you are required to or not. Who knows

what the first reroofers covered up. Also, the bumpy surface of two layers of crumbling shingles makes it almost impossible to put down a third roof well. Three layers of shingles might not fit under existing chimney flashing or under the siding of adjacent dormer walls. In some places curled and buckled shingles will keep nails from penetrating the sheathing; in other places you'll strike unseen voids and split or tear the new shingles as you nail. Also, it's best to rip off old wood shingles—and even the skip sheathing if it's damaged—and start from a clean deck. If you don't, you could cover up rot when you lay down your new asphalt shingles.

And don't forget that each layer of roofing adds some weight: approximately 2.35 lb. per sq. ft., or 235 lb. per square. (A square is 100 sq. ft.) Each buried layer is an extra burden on the shoulders of an old house. Another consideration is the fire hazard of adding a third layer of petroleum-based material to a wood-frame structure. Most fiberglass shingles are class-A fire-resistant; however, no shingle is completely fireproof.

To sum it all up, if I find underlying structural damage, the old roof has to come off. If there are two or more layers of roofing, they must come off in any case. In general, ripping off the old roof makes for not only a more watertight, longer-lasting roof but also a better-looking one. And it makes putting on the new roof easier.

Prepare for tearoff—Before tearing off a roof, we protect the grounds with large plastic tarps. Any windows, doorways, shrubbery, etc. that might get damaged are also covered with tarps. Whenever possible, we nail a tarp to the fascia board or soffit and let it hang to the ground to protect the house from falling debris and dirt (photo facing page). There are always extra tarps and rolls of roofing felt ready in case the weather unexpectedly changes. (You might pay attention to the weather forecast.) Inside the house, we

Spiking the cleat. A low-tech but effective foothold when working on a particularly steep roof is a 2x4 cleat spiked over the old roofing with 12d duplex nails, which are easy to pull out. These 2x4 cleats are spiked into the roof vertically every few feet, and a row of them continues unbroken across the roof.

Roof bracket. On a steep roof, these brackets provide stable workstations. They are nailed through the roofing into rafters, and a plank is nailed to the brackets through holes at the front of the bracket. Because of the teardrop shape of their nail slots, roof brackets are easily removed by tapping them upward.

Setting up ladder brackets. With extension ladders tied off at the fascia, ladder brackets are hooked onto the ladders about 5 ft. below the eaves. Scaffolding planks—not framing lumber, which is too springy—span the brackets to give you a place to work from when you begin roofing.

cover the attic floor with plastic because debris and dirt fall through the cracks in the sheathing—particularly with board sheathing but even with plywood—and mess up the attic.

Picture windows and French doors are sealed with plywood. On the roof, we protect skylights by taping ¼-in. plywood or heavy cardboard over the glass. If we remove a vent or find or cause any other hole in the roof, a piece of plywood is nailed over the hole temporarily so that no one inadvertently steps through it.

Most important, before tearing off the roofing, we decide where to throw it on the ground and how to get rid of the stuff when we're done. On smaller jobs, we use the dump truck. Around here it costs $80 per ton to dump debris at the transfer station.

In many places you can't bring old shingles to the local dump. Construction debris must be transported to a transfer station that sorts debris into recyclable and nonrecyclable material.

Recycling of roofing material is in its infancy. According to Stuart Laughlin of the Bird Corporation, which manufactures shingles, defective new shingles are now being recycled as a cold-mix asphalt base for roads. The problem with old shingles is that no one has found a way to separate the nails. Bird is experimenting with an elec-

tromagnetic process, but for now, unless you can pull every nail, you won't be able to recycle the old roofing.

On bigger jobs, we rent 1 yd. of dumpster for every square of roofing. Our dumpsters come from the private contractor that runs the transfer station. A 30-yd. dumpster costs $410, which includes delivery to the site and pickup when it's full. If the dumpster is filled more than once, it costs an extra $65 per ton of material. There isn't much choice when it comes to finding a place for the dumpster. It's got to be accessible to both the roofers and the carting company, so it usually ends up on the driveway near the house. The driveway's good because a dumpster will sink into the ground. To minimize driveway damage, we have the dumpster set on pieces of plywood.

Old shingles will ruin a patch of lawn in a day, so we try to clean the grounds as we go—the safest, most-efficient way to work. If you leave the stuff on the ground for a while, at least throw it onto a tarp, cover it each night and get it into the truck or dumpster before it becomes a giant, rain-soaked, nail-studded pile.

As long as the roof isn't skip-sheathed, the only difference between tearing off wood shingles and tearing off asphalt shingles is that the transfer station requires wood shingles to be separated

from roofing felt for purposes of disposal. Regardless of the material, we tear off old roofs in sections, removing only as much as we can make watertight before the end of the day. Roofs with dormers, vents, skylights, etc. take longer to tear off and make watertight than, say, a straight gable, so pace yourself accordingly.

Getting on the roof and staying there—First we set up staging from the ground to the eaves. This staging may be metal scaffolding, and in places where shrubs close to the house make it hard to set up scaffolding, we set two 2x12 planks on ladder brackets (left photo, above). We tie off the tops of the ladders to keep a sliding clump of shingles or a roofer stepping from the ladder to the roof from knocking the ladder sideways.

Once we're on the roof, we nail pairs of roof brackets vertically (adjustable, triangular metal supports for planks) onto the roof every 6 ft. or so (bottom right photo, above). The highest brackets are positioned about 6 ft. below the roof's peak, and all the brackets are nailed through the sheathing into the rafters. Planks span each pair of brackets (we use real planks: full 2x12 spruce, not framing lumber, which is too springy), and a roofing nail driven through each bracket into the planks prevents them from twisting or lifting up.

On a particularly steep roof (or when we run out of brackets), we nail 2x4 cleats at the same or smaller intervals up the roof to stand on (top right photo, facing page). We use 12d duplex scaffold nails, which have a double head, so they're easy to pull out. We always make a continuous row of cleats across the roof with no gaps between them. It is a long first step to the ground.

Safety on the roof—The first rule of safety on any roof is don't fall off. I wear sneakers. Admittedly, they don't protect against the other hazard—stepping on nails—but I've found that good cross-trainers provide the best traction on a slippery roof. You may prefer thick-soled boots, but remember that boots are a lot rougher on new asphalt shingles than sneakers are.

One question should govern how you work on a roof: Are you happy there? If you're not happy, you must reevaluate the setup. The most secure staging won't keep a person who is afraid of slipping from doing just that. Trust your fear and make the workstation *too* safe. Make yourself happy there.

On a typical job with the Wright Brothers' roofing crew, two rippers start work at the peak on the uppermost plank, one person is on the staging below them, and one man patrols the ground. The two rippers tear off the roofing. The person below moves back and forth, taking debris out of the rippers' way, keeping their staging clear and sending the material along to the ground. The ground man, always alert to what's falling from the sky, moves the pile to the dumpster. The ground man should wear thick-soled boots rather than sneakers. He will be wading in shingles. He should also wear a hard hat. Everyone should wear gloves.

Tooling up for the job—When tearing off shingles, we carry 16-oz. straight-claw hammers. The easier the material comes off, the less we use the hammers. But they will be ready to use on nails that remain in the sheathing and for patches of cemented shingles that can't otherwise be dislodged. The straight claw is the roofer's hammer. The story you hear now and again of a roofer who began sliding off the roof and stopped his fall by swinging the claw end of the hammer into the sheathing and hanging on is true.

The ripping shovel, or shingle ripper, is the main tool (top photo, this page) for removing shingles. It's similar to a long-handled spade, except the blade is completely flat with large serrations across the tip, and the handle is steeply angled to give extra leverage for prying up the shingles. You don't remove shingles the way you would shovel dirt. You drive the ripping shovel under as many courses of shingles as possible, and by pushing down on the handle, you spring the shingles loose from the sheathing. The roofing nails hook in the blade's serrations just as they would in the claw of a hammer and are pried up along with the shingles.

Most roofing nails come up with the shingles. The recalcitrant nails are removed later in the final cleanup before papering in the roof (the process of rolling out and fastening felt paper to the sheathing).

A roofer and a ripper. The main tool for removing shingles from the roof is a shingle ripper, or ripping shovel. Its flat, serrated blade gets under shingles and around roofing nails. The handle is angled steeply so that when you push down on it, you loosen both the shingles and the nails.

Almost as good as the ripping shovel is a regular garden pitchfork. On an old roof with board sheathing, the prongs tend to stab into the seams between boards, but if you keep the pitchfork about parallel to the roof as you drive the fork under the shingles, it will work fine.

That's about it for tools. Pneumatic and gas-powered ripping tools are available, but I've never used them. A ripping bar, a heavy scraping tool for removing tar-and-gravel roofs, may be used to get up cement around chimneys, but usually cement clings to shingles, and leftover clumps of it can be knocked off with a hammer. Later, I'll need a flat pry bar, but for now it's just pitchforks and ripping shovels.

Ripping and tearing—There are no real tricks to ripping. Once the first patch of bare sheathing has been exposed, we work out in all directions from it. Sometimes a section will come up easier by prying from below; other times we can pry a section off by standing above it. I get the shovel under as many layers as possible and try to spring the nails loose. Just digging and ripping won't do the job; small clumps of material will come up, but most of the shingles will stay nailed down. I save a lot of work by prying as many nails loose as possible and

Finesse the flashing. Be careful when tearing off shingles around chimneys, vents or dormers because you can damage the flashing or the structure. So use a shingle ripper and get close, but pry up the flashing and the shingles around it with a flat pry bar. If the flashing is in decent shape, you might want to save it.

Roll out the paper. You can save a little extra work and keep from walking on the tar paper by papering from the peak down. Before nailing the bottom edge, tuck the lower course under the upper and fasten the whole shebang with roofing nails and tins, which hold the paper better than staples do.

releasing as many layers of shingles at the same time as possible.

Then again, we're careful not to pull up too much at once. It can happen that, once started, the whole roof begins to come up in a vast sheet. I break it into sections about 3-ft. square so that when I throw the shingles down, I won't heave myself off the roof with them.

Loose shingles are slippery and dangerous. I remove as much as I can safely reach from along the roof brackets or the cleats and clean up or have my cleanup person finish that section while I move to the next area. Even if I'm momentarily knee-deep in shingles, I always stand on a clean plank or a solid part of the roof—never on the loose shingles. I keep a push broom handy to sweep the roof and the staging. The granules from the shingles are particularly treacherous.

Working around flashing—It's best to leave chimneys, vents, skylights, roof-to-wall intersections and valleys until last. Here is where most damage to the underlying structure probably is, and it's where most damage can occur if you're not careful about tearing off the roofing here.

Using a pry bar and a hammer, I pry up these last remaining shingles (bottom photo, p. 17). Around old vent pipes, there's an iron-ring weather seal. I break it off with a few hammer blows and slip an aluminum-flanged flexible rubber boot over the vent pipe. The flange is nailed along the top, and a course of shingles is tucked under the bottom flange. Around chimneys, I very carefully chisel away any old roofing cement with the pry bar and the hammer and pry up the chimney flashing and counterflashing—

without removing or damaging either one. The counterflashing is often made of lead and tears easily. It is set into the mortar between the courses of brick as the chimney is being erected, and its replacement is a job for a mason. I bend the pieces away from the roof without tearing them or disturbing their positions in the chimney.

I might simply remove step flashing, which is woven into courses of roofing at walls and skylights, but I always think of the consequences: Can I put new step flashing in without removing courses of siding? If the flashing is undamaged—and it doesn't leak, and the courses of the new roof will line up properly with the old flashing—I bend the bottom edges of the step flashing up a bit to clean under it, then I weave the new shingles into it when I reroof.

It's hard to detect small cracks in old valley flashing. Even when it appears intact, it's not worth leaving only to find later that the new roof leaks and to wonder if the flashing should have been replaced when the roof was open. So we remove and replace all valley flashing. Occasionally we leave the original flashing and install a new, wider piece over it. In any case, don't walk on that new flashing because walking on it will cause leaks.

Now we're down to the sheathing. We sweep the roof clean and go over it carefully, pulling up or pounding in all remaining nails.

We stop tearing off the old stuff when we've cleared a manageable section: a section that can be repaired and made watertight by the end of the day. On a small gable we may be able to tear off and make watertight half of the roof in one day. On a large roof, where staging must be

moved often, we reroof the section we've torn off before dismantling staging, brackets and planks so that we won't have to remount them later. It's possible to cover the whole roof with tarps, removing them each day to work and replacing them each night, but we prefer to complete a section, paper it in and have it ready to reroof before moving on.

Repairing damage—A rotted piece of sheathing, even if it's only a small section, should be replaced with a new piece that spans at least three rafters. I find the rafters, remove the nails from the sheathing with a nail puller, set the circular-saw depth to just beyond the thickness of the sheathing and cut it at the centerline of the two end rafters. Then I renail the old sheathing at the cutlines and put in the replacement piece.

Old board sheathing is commonly $\frac{7}{8}$ in. or a full 1 in. thick. To replace a few rotted boards, I use $\frac{5}{4}$-in. rough spruce ledger board if it's available. Where there are lots of boards to replace, I use $\frac{5}{8}$-in. exterior plywood and shim the rafters with $\frac{1}{4}$-in. or $\frac{3}{8}$-in. lattice molding, available in various widths at any lumberyard, to bring the plywood level.

If the original sheathing has shrunk, and there are spaces between boards, or if the sheathing is sagging between rafters, I often put a second layer of sheathing over the first. The second layer is usually $\frac{3}{8}$-in. plywood staggered and nailed into the rafters. In particular, I avoid joining the plywood in the same pattern as the old sheathing. Staggering the joints will strengthen an old roof and provide an even, secure nailing surface for the new shingles.

Thwarting ice dams. To get a watertight seal that protects against leaks caused by ice dams, use Ice & Water Shield. It sticks to the sheathing because it's coated on one side with adhesive. Put a single layer of it at the edge of the roof over the bottom drip edge before papering in. Some roofers will even run Ice & Water Shield in valleys for added leak prevention.

Dragging the magnet. You'll have fun picking up nails and all kinds of junk with this sporty magnetic bar. Pull it around the yard, in gardens, over the driveway and the street.

Once in a while, we're asked to put asphalt shingles over an old wood-shingle, skip-sheathed roof. We tear off the wood shingles because it's difficult to apply new roofing over them. Then, if the skip sheathing is in good shape, we install a new plywood or OSB roof deck over it, adding molding at the rakes and the fascia to cover the gap. If the skip sheathing has deteriorated, we tear it off down to the rafters and put down a new roof deck.

If the rafter ends are rotted, we remove enough sheathing to scab the new rafters to the old where the old ones are solid. Opening the roof allows us to scab on new rafter tails without disturbing the interior of the house. We may even be able to leave soffit and trim if they are solid. The general rule for scabbing new overhanging rafters to old ones is to extend the new rafter above its bird's mouth (or bearing point) on the exterior wall plate twice the length of its overhang. For example, a 2-ft. overhang requires a 6-ft. length of rafter, 4 ft. of which runs inside the house. If only the last inch or two of the rafters has rotted, this rule doesn't apply; we just scab new pieces to the solid sections of the overhanging members and then cut off the rotted portions.

Papering in—We paper in and prepare for the new roof, even if it won't be shingled immediately. We paper in from the peak down, making whatever's exposed watertight and then continue tearing off the lower part of the roof. Using 15-lb. roofing felt, we paper right over the peak. On the other side of the peak, we put strapping over the paper and nail it down onto the old shingles. Strapping holds the paper securely and is easily removed when we start tearing off the other side.

Because we start at the peak, we nail the felt at the top and the middle only. When we move down the roof, we simply slide the next underlying course of felt under the upper one and tin and nail through both (photo facing page).

We secure the felt with roofing nails and tins. The tins, called buttons in some areas, are aluminum disks that you nail to the felt. Tins secure the felt to the roof much better than staples or nails alone would. Most lumberyards carry either tins or nails with a big, square washer already attached. Properly tinned and nailed, the felt will remain secure against rain and wind until we shingle. (But don't walk on the felt. You won't stay on the roof long if you do. After all, you're walking on impregnated paper that tears easily.)

Across the bottom 3 ft. of the roof, we put a layer of Ice & Water Shield (W. R. Grace and Co., P. O. Box 620009, Atlanta, Ga. 30362; 800-444-6459)—a polyethylene material coated on one side with mastic roofing cement (left photo, above). The shield adheres to the roof sheathing and seals it against water backing up from an ice dam. It does not prevent either an ice dam or the backup, but it does form a watertight barrier. Ice & Water Shield comes in 3-ft. by 75-ft. rolls and costs about $75 per roll.

If the roof will have woven valleys with no flashing, we put a length of Ice & Water Shield in the valley before reroofing. Most of the time, we flash valleys first, put a length of Ice & Water Shield along each edge of the flashing and then install the roofing to extend a bit past the Ice & Water Shield.

Ice & Water Shield sticks to the sheathing, to you and to itself. It's a little like working with a giant role of electrical tape. The shield goes on over the bottom drip edge but under the new valley flashing. The material is backed with brown waxed paper to prevent it from sticking to itself on the roll. Two workers can handle the material better than one can. You first peel back about 6 ft. of the paper. Then, with one person handling the roll, the other carefully places the exposed material parallel to the bottom of the drip edge, to which it will immediately stick. Once the shield is secure, the rest of the roll can be peeled from the backing and slowly rolled across the length of the roof.

Final cleanup—Even though we'll have to clean up all over again when we finish shingling, we clean up thoroughly before reroofing, then clean out the gutters. They will be full, and if they're aluminum, they'll be ready to buckle. We rake out the bushes and the yard and clean the driveway. For now, we leave the plastic in the attic because hammering and walking on the roof will certainly shake down more dirt. Then we break out the rolling magnetic bar (right photo, above). Mine is simply a 2-ft. long bar magnet attached to a rope. (A similar tool is available from Haase Industries, Inc., P. O. Box 450, Lake Oswego, Ore. 97034; 800-547-7033.) I drag it along the driveway and around the yard. The bar won't pick up most types of flashing, but it's great at collecting nails. □

Jack LeVert is a carpenter and author living in Natick, Mass. Photos by Rich Ziegner.

Roof Shingling

With only a few rules to follow, putting on a wood roof can be relaxing work with pleasant materials

by Bob Syvanen

Shingling is one of my favorite tasks in building houses. Even though roofers may be a little faster, I like to do this work myself. Shingling is the kind of job that requires little calculating and a minimum of physical effort. You can think of other things as you work. You don't have to manipulate unwieldy boards or carry heavy loads. Shingle nails are fairly short, so swinging the hammer is easy on the arm; and if you have the time to invest in the old-fashioned methods of making hips, ridges and valleys, there's just enough cutting and fitting to make the job interesting.

Wood shingles are typically three or four times as expensive as asphalt shingles, but they give a roof a texture and color that you can't get with petroleum products. A wood roof is also much cooler in the summer, and will last nearly twice as long as one covered with conventional asphalt shingles. The only major disadvantage to wood shingles is their flammability; but chemical treatments, along with spark arrestors on fireplace chimneys, can minimize this liability.

In the past, shingles were commonly made of cypress, cedar, pine or redwood. My favorite is cypress, although red cedar is what's most available these days. It too is excellent for roofs because the natural oil in the wood encourages water to run off instead of soaking in, and it helps prevent the shingles from splitting despite wide fluctuations in humidity and temperature year after year.

Wood shingles are a delight to work because they're already cut to length and thickness from the best part of the tree, the heartwood. Shingles are sawn flat on both faces, which distinguishes them from shakes, which are split out along the grain. Wood shingles come in lengths of 16 in., 18 in. or 24 in., and taper along their length. The exposed ends, called butts, are uniformly thick for each length category of shingles. This measurement is always given in a cumulative form—16-in. shingles, for instance, always have butts that are 5/2. This means that five shingle butts will add up to 2 in.

Tools—Tools for shingling are few and simple. I have put on many shingles using a hammer and a sharp utility knife. Most pros use a lathing or shingling hatchet. A shingling hatchet has an adjustable exposure gauge on the blade; however, a mark on your hammer handle works almost as well. Two good features of the hatchet are the textured face on the crown, and the hatchet blade itself. The mill face or waffle head is less likely to glance off a nail onto a waiting finger, especially when the head of the hammer strikes a blob of zinc that hot-dipped shingle nails often have. The sharp blade and heel of the hatchet are useful in squaring shingles, and in trimming hips and rakes. I also use a block plane to trim hip, valley and ridge shingles for final fit.

I prefer to keep my nails in a canvas apron at my waist, but others like leather nailbags hung off a belt. Production roofers use a stripper, a small, open aluminum box that straps to their chest. It has slots that allow the points of the nails to drop into line when it's loaded with a handful of nails and shaken back and forth.

Although it's possible to work from a ladder, proper staging or scaffolding is a big help when starting a roof. Wall brackets (drawing, facing page, left) are my first choice. You can make them from 2x4s, or rent or buy the sturdier steel ones. They should be attached to the wall studs at a comfortable working height below the starter course on the eave. Some brackets will accommodate nearly 30 in. of scaffolding planks, but two 2x10s battened together make

---

Figuring materials

You can usually buy shingles in three grades. Always use No. 1, the best grade for a roof. No 1. shingles are 100% clear heartwood and edge grain. No. 2s and 3s have more sapwood, knots and flat grain. They're okay for outbuildings where the life expectancy of the structure is shorter, or for starter coursing, shim stock and sidewall shingling.

In order to figure how many shingles you will need, you must first know what exposure you are going to use. Exposure is the measurement of how much of the shingle shows on each course. The longer the shingle, the greater the possible exposure. Maximum exposures are also determined in part by the pitch of the roof. As shown in the chart below, the flatter the pitch, the less shingle that can be left to the weather.

Using these maximum exposures, all pitches of 5-in-12 and up will give triple coverage. With lesser pitches, successive courses of shingles will overlap each other four times, so you get quadruple coverage. Wood shingles shouldn't be used on pitches lower than 3-in-12.

Maximum exposure (in.)			
Roof pitch	Shingle length (in.)		
	16	18	24
5-in-12 and up	5	5½	7½
4-in-12	4½	5	6¾
3-in-12	3¾	4¼	5¾

When the maximum exposure is used on any length shingle, four bundles of shingles will cover 100 sq. ft. of roof, or one square. To figure your roofing material, multiply the length of your roof by its width and divide by 100. This will give you the number of squares. For starter courses, add one extra bundle of shingles for each 60 lineal feet of eave. A starter course is the first course of shingles, which is doubled to provide a layer of protection at the joints between shingles. In some cases, the starter course is even tripled. For valleys, figure one extra bundle for every 25 ft., and the same for hips. If there is a hip and a valley, figure some of the waste from the valley to be used on the hip. If you are going to use manufactured hip and ridge shingles, a bundle will cover about 17 lineal feet.

When you calculate shingle quantity, figure nails and flashing, and order them at the same time. For 16-in. shingles on a 5-in-12 or steeper pitch, use 3d galvanized shingle nails and figure 2 lb. per square. For 24-in. shingles, get 4d nails, with 5d nails or bigger for re-roofing jobs when you're nailing through other shingles. Hip and ridge caps need nails two sizes larger than the shingle nails used in the field (on the roof slope), because you will be nailing through extra thicknesses.

If you are near a good-sized city, roofing-materials suppliers are fairly common. Their prices are often more reasonable than lumberyard prices, and their inventory is only for roofers, so you're more likely to get the flashings and nails that you need. Ask for the price per square on the shingles that you want, and be prepared to give a figure of how many squares you'll need. Another advantage of buying from a roofing supplier is that many of them deliver on lift-bed trucks and load the roof with the shingles, saving your back and a lot of time walking up and down a ladder. —*Bob Syvanen*

a nice working platform. Make sure the battens (or cleats) extend back beyond the planks to the wall to prevent them from shifting on the brackets inward under the eaves. Scaffolding planks and steel ladder brackets hung from the rungs of two straight extension ladders placed against the siding will also work nicely. Once up on a roof over a 4-in-12 pitch, I use roofing brackets, but I'll get to these later.

Preparation—One decision you make before shingling is what sheathing to use. Where there is no wind-blown snow, or where the weather is humid and wet much of the time, an open slat roof with spaced sheathing is a good choice. Shingles are laid on top of it without roofing felt, so air can circulate freely on the underside of the shingles. The spacing of this sheathing is important. For a 5-in. exposure, the 1x4 sheathing should also be spaced 5 in. o.c. (drawing, below right). The shingle tips should lap over the sheathing at least 1½ in., with two boards butted together at the eaves and ridge for proper nailing of starters and ridge caps.

In snow country, the solid-sheathed roof is best. I have stripped many roofs with solid sheathing and found that they have held up very well. I use CDX plywood and cover it completely with 15-lb. felt. If you are using felt, lay only as much as you need for a day's work. Morning moisture will wrinkle the paper and make shingling difficult. Along the eaves I use 36-in. wide 30-lb. felt. If ice-damming is a particular problem in your area, you can trowel on

Using a shingler's hatchet, Syvanen squares up a red cedar shingle, right. The 1x3 roof stick he is cutting on is a gauge for laying straight courses with the correct exposure. The shingles on the top course at the left of this photo have been butted to this gauge and nailed, each one with two 3d shingle nails.

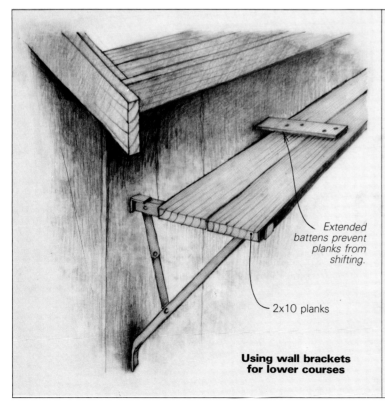

Extended battens prevent planks from shifting.

2x10 planks

Using wall brackets for lower courses

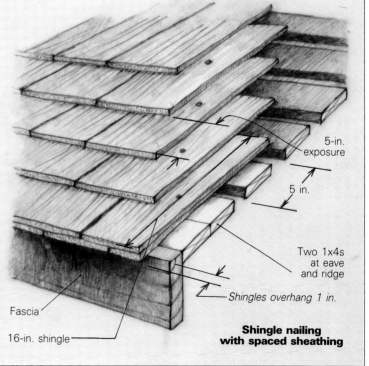

5-in. exposure

5 in.

Two 1x4s at eave and ridge

Shingles overhang 1 in.

Fascia

16-in. shingle

Shingle nailing with spaced sheathing

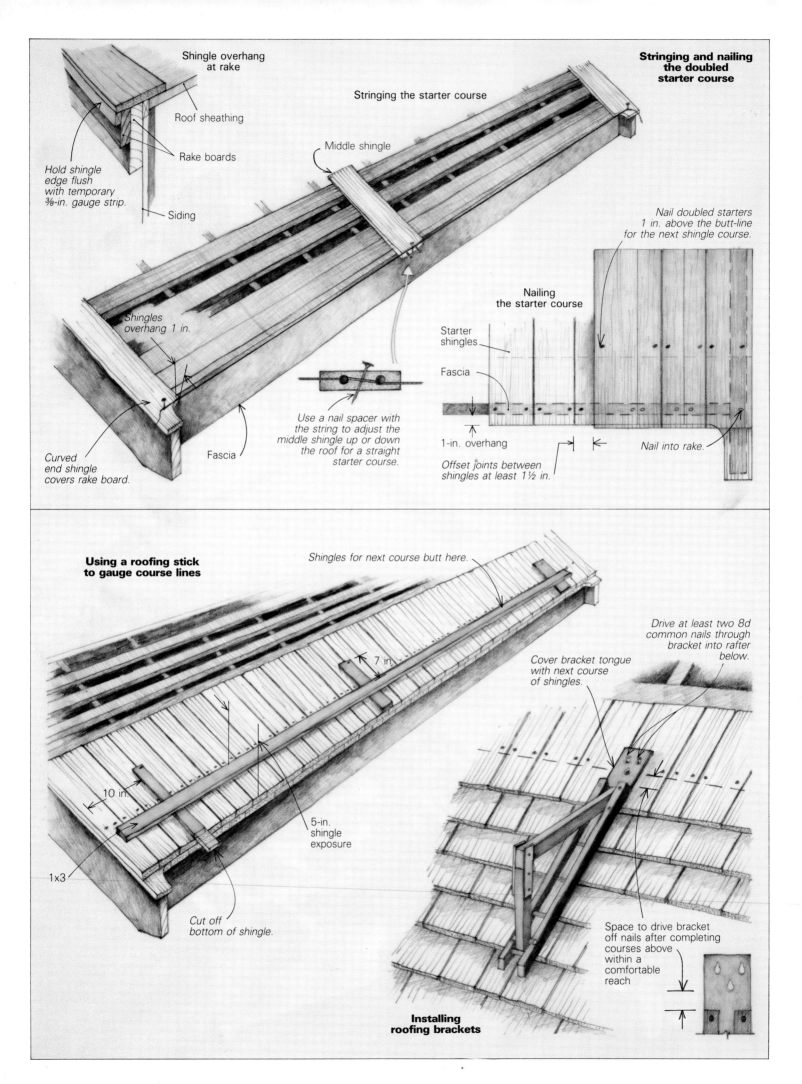

Shingle overhang at rake

Roof sheathing

Rake boards

Hold shingle edge flush with temporary ⅜-in. gauge strip.

Siding

Stringing the starter course

Middle shingle

Stringing and nailing the doubled starter course

Nail doubled starters 1 in. above the butt-line for the next shingle course.

Shingles overhang 1 in.

Nailing the starter course

Starter shingles

Fascia

Curved end shingle covers rake board.

Fascia

Use a nail spacer with the string to adjust the middle shingle up or down the roof for a straight starter course.

1-in. overhang

Offset joints between shingles at least 1½ in.

Nail into rake.

Using a roofing stick to gauge course lines

Shingles for next course butt here.

7 in.

Drive at least two 8d common nails through bracket into rafter below.

Cover bracket tongue with next course of shingles.

10 in.

5-in. shingle exposure

1x3

Cut off bottom of shingle.

Space to drive bracket off nails after completing courses above within a comfortable reach

Installing roofing brackets

Illustrations: Frances Ashforth

When using roof brackets, or jacks, squatting is usually the most comfortable working position. Once you have shingled up the roof to a point where you are working at full arm extension, nail a new line of brackets to the rafters higher up the roof. These brackets are removed when the roof is completed by tapping them at the bottom. The 8d common nails used to secure each bracket remain under the shingles.

a layer of roof mastic over the 30-lb. felt at the eaves, and then lay on another run of felt over this for a self-sealing membrane.

The shingles will need to overhang both the rake and eaves. For the rake overhang, cut some temporary gauge strips ⅜ in. by 1 in. or so, and tack them on the rake board with 4d nails (drawing, facing page, top left). You can then hold your shingles flush with the outside of this gauge board for a ⅜-in. overhang.

Starter course—The first course along the eave is doubled. I like to extend the end shingle of this starter course over the rake board or gutter if there is one. Water runoff will wear away the top of the exposed piece of rake board if it is not covered. This extended shingle can be straight or curved. I like it curved. I use a coping saw and cut several at one time.

To line the starter course, nail the curved end shingles in place. These shingles will be flush with the rake gauge strips and will overhang the fascia on the eave 1 in. Tack a shingle in the middle of the run with the same 1-in. overhang. Because the fascia probably won't be truly straight, the middle shingle will have to be adjusted up or down to straighten the starter course. Do this by stretching a string from one end shingle to the other on nails spotted at the line of the 1-in. overhang.

To keep from butting the shingles directly to the string and introducing cumulative error by pushing against it, fix the string away from the line of the starter course by the diameter of a nail. Then use a loose nail to gauge each starter shingle to the string as you nail it. Begin by adjusting the middle shingle to the string in this way. The drawings at the top of the facing page show how to get the string right.

Once the middle shingle is nailed down, fill in the rest of the starters, nailing them all into the fascia. Nail the end shingles into the rake board (drawing, facing page, top right). Angle the nails away a little to make sure that they don't poke through the face of this trim.

On top of the starters, nail the double starter shingles, making sure that the gaps between them are spaced at least 1½ in. from the gaps in the row below. The shingles in this course can overhang the starters by about ⅛ in., or sit flush. They should be nailed about 1 in. above the butt line for the next course; for a 5-in. exposure, nail about 6 in. up on the shingles.

Roof shingles get a lot of water dumped on them, and they will buckle if placed too close. I just eyeball the distance between edges, but any spacing from ⅛ in. to ¼ in. is fine. The joints between shingles should be offset from the joints in the course directly below by at least 1½ in., and offset from the gaps in the course below that one by at least 1 in.

Roofing sticks and brackets—To lay the second and all successive courses, you'll need a method of gauging the exposure and keeping the courses straight. One way is to use the gauge on your hatchet or a mark on your hammer handle. I prefer to make and use roofing sticks to align an entire course before I have to reposition them. Use as many of these sticks as it takes to span the roof. To make one, take a long, straight length of 1x3 and nail a 2-in. wide shingle to it at right angles every 8 ft. or so (drawing, facing page, bottom left). Then saw off the butt of the shingle so that the distance from this cutoff edge to the top edge of the 1x3 equals your single exposure. To use the roofing sticks, line up the shingle butts on the gauge stick with the butts of the first course. Tack the gauge to the roof with a nail in the upper corner of each gauge shingle. By butting the shingle ends down against the top edge of the 1x3, you get uniform exposure from one course to the next. On a calm day, you can lay in quite a few shingles before nailing them down—a nice feature of this system.

When you reach one of the shingle tips that is part of the roofing stick, just fit the shingle without nailing it. Tuck it under the butt of an adjacent shingle for safekeeping and continue down the course. Then when you've completed the full course, tap the roof sticks free and nail down the individual shingles you left out.

On even a moderate pitch, you'll need roof brackets to work safely, comfortably and at a good pace. These are adjustable jacks that are nailed to the roof to support a scaffolding plank (drawing, facing page, bottom right). Most lumberyards carry roof brackets, and they can also be rented. The ones I use are oak, but metal ones are more common. The first set can be put on after you've shingled up five or six courses and can't reach across the eave from the ladder or scaffolding.

Make sure your roof brackets are nailed into rafters, and use two 8d galvanized nails per bracket. Most brackets will take a single 2x10 plank. Twelve-foot planks are ideal when you

use a bracket at each end. Fourteen-footers will do. You can shingle right over the tongue of the bracket where it attaches to the roof, because the nails that hold it will be left under the shingles when it's removed. When you install the brackets, make sure there's enough space between the top of the bracket and the shingle butts to allow you to remove the bracket easily, by driving its bottom toward the ridge, and lifting the tongue off the nails.

Do's and dont's—Face-grain shingles have a right and wrong side up. Make sure to place the pith side of the shingle (the face that was closer to the center of the tree) down to prevent cupping. Some shingles are cupped or curled at their butts. I put the cup down for better runoff and appearance. I reject hard shingles because they split when nailed and curl up when the sun hits them.

Red cedar shingles can be very brittle, so hold back on that final hammer stroke or the shingle may crack. The nail head should sit on top of the shingle, and not be driven flush. Each shingle, no matter how wide, should get just two nails. If you have a hip or valley to shingle, save the bedsheets (shingles 9 in. or wider) when you're roofing in the field.

If you use a wide shingle in a regular course, score it down the middle with a hatchet or utility knife to control the inevitable splitting. Treat it as two separate shingles, offsetting the scored line from other joints, and nailing as you would two shingles. If you suspect that a shingle is checking, bend it into a slight arch with your hands. If it has a clean crack, use it as two shingles. If it shows several cracks, throw it out.

When you are shingling in the field, don't get so tightly focused on the work that you forget to check every once in a while that the coursing is even and straight. Do this by measuring up from the eave, and adjust if necessary. You may have to snap a chalkline occasionally to straighten out a course.

Once you've completed the roof, spend a few minutes looking for splits that you missed that

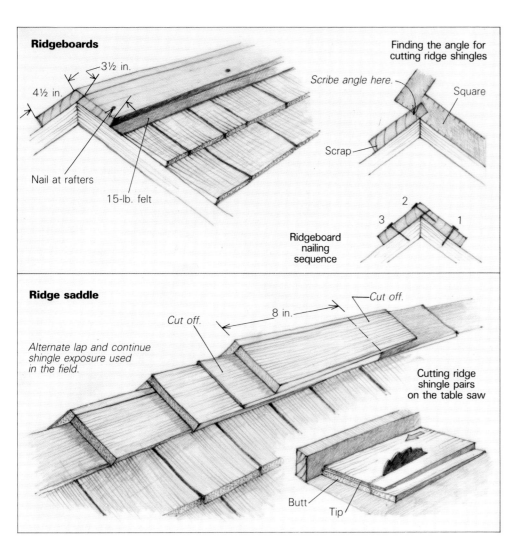

Ridgeboards

3½ in.

4½ in.

Nail at rafters

15-lb. felt

Finding the angle for cutting ridge shingles

Scribe angle here.

Square

Scrap

Ridgeboard nailing sequence

3 2 1

Ridge saddle

Alternate lap and continue shingle exposure used in the field.

Cut off.

Cut off.

8 in.

Cutting ridge shingle pairs on the table saw

Butt

Tip

are not sufficiently offset from joints between other shingles. Cut some 2-in. by 8-in. strips of flashing out of zinc, aluminum or copper, and slip them under the splits.

Ridges—Make sure that the courses are running parallel with the ridge before getting too close to the top. Since a course of 3 in. or less doesn't look good at the ridge, measure up from the course you are working on to the point where the ridgeboard or ridge shingles will come, and adjust the exposure slightly so that the courses work out. If you are using ridgeboards, staple 15-lb. felt over the ridge for the full length of the roof. The felt should be an inch or so wider than the ridgeboard, and will make painting the final coats on the ridgeboard easier. Make your ridgeboards from 1x material, and bevel the top edge of each board at an angle that will let them butt together, and form a perfect peak. Because the boards butt at the ridge rather than miter, one side will be narrower than the other (drawing, top left).

To find the bevel angle for your roof on the table saw, take a pencil, a scrap of wood and a square up to the ridge. Lay the scrap flat on the roof on one side of the ridge so that part of the scrap projects over the peak, as shown in the drawing. Then lay the square flat on the other side of the roof so that it projects past the scrap. Scribe a line onto the wood by holding the pencil against the square. By tilting your table-saw blade to this angle, you will be able to rip both the narrow and wide ridgeboards.

To install the ridgeboards, snap a chalkline

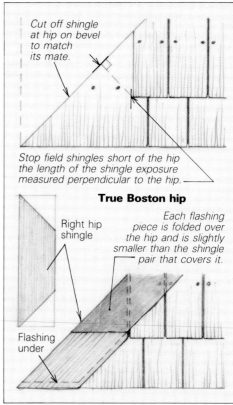

Cut off shingle at hip on bevel to match its mate.

Stop field shingles short of the hip the length of the shingle exposure measured perpendicular to the hip.

True Boston hip

Right hip shingle

Each flashing piece is folded over the hip and is slightly smaller than the shingle pair that covers it.

Flashing under

A true Boston hip, shown at left in a model built by Syvanen, uses the hip shingles to complete each course. The usual way is to superimpose a line of ridge shingles on the hip. The butts of Boston hip shingles can run parallel to the eaves, as shown on the right side of the model, or perpendicular to the hip (left side).

the length of the roof where the lower edge of the narrow ridgeboard will be nailed. Using 8d common galvanized nails, attach the narrow ridgeboard to the roof along the chalkline. Be sure to nail into the rafters. Now nail the wide ridgeboard to the narrow one. This will give a nice straight line at the peak. Lastly, nail the lower edge of the wide ridgeboard into the rafters. Push down on the ridge pieces to force the narrow board against the roof. You might have to stand on the ridgeboards to do this. After the final coat of paint is on, trim off the 15-lb. felt with a sharp utility knife.

You can also use shingles for the ridges—either the factory-made kind that come already stapled together as units, or your own, cut on a table saw. To make your own, set the blade-to-fence distance for the width of your exposure, and set the angle of the blade as explained for cutting ridgeboards. Saw the shingles in alternating, stacked pairs. The bottom shingle in the first pair can have the butt facing the front of the saw table, and the top shingle with the tip doing the same. The next pair of shingles should be reversed so that the tip is on the bottom, and the butt is on top. This will give you shingle pairs whose laps alternate from one side of the ridge to the other.

To install the ridge shingles, staple a strip of 30-lb. felt down the full length of the ridge. In this case, the paper should be narrower than the ridge unit. Nail the ridge pieces along a snapped chalkline using a double course as a starter. Alternate the pairs and use the same shingle exposure you have on the rest of the roof. Again, the nails should be about 1 in. under the butt of the next ridge shingle. When you reach the center, make a saddle by cutting the tips off two pairs of ridge shingles so that 8 in. remain on each (drawing, facing page, center left). Then nail them down on top of each other with their butts facing the ends of the ridge.

Hips—You can buy factory-assembled ridge units for hips, or make your own. Most folks just staple down a run of 30-lb. felt over the hip and nail the units along a snapped chalkline or temporary guide boards. Since water drains away from a hip, this method keeps the rain out. But when I have the time, I like to cut and fit a true Boston hip the way the old-timers did (photo and drawing, facing page, bottom). It weaves the hip shingles and flashing right into the courses in the field for a weathertight fit that doesn't look added on, like standard hip units. It looks tough to do, but it's not really.

The key to working a true Boston hip is to stop the shingles in the field the same distance from the hip on each course. This makes the hip shingles uniform, allowing you to cut them on the ground. To find the point on a course line where the butt of the last shingle should end, measure the length of the shingle exposure you are using in the field (such as 5 in.), on a line that runs perpendicular to the hip. You will have to move this 5-in. line, marked on a tape measure or square, up or down the roof (keeping perpendicular to the hip) until it fits between the hip and the course line. Then

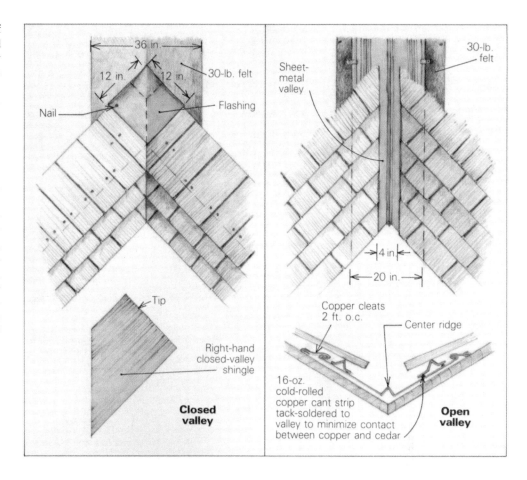

mark the intersection on the course line and fill in the shingles in the field to that point.

You will need to make right and left hip shingles to form pairs. The tip of each hip shingle needs to be trimmed to fit into the space left for it by the last shingle in its course and the one above it. The butts can be left perpendicular to the hip, or cut parallel to the eave. The long side of the shingle also needs to be cut. This should be a bevel that alternates lapping its mate on the other side of the hip. You can cut these bevels on a table saw or use a shingling hatchet on the roof.

Flashing pieces, which are used under every hip shingle, should be cut out with snips on the ground. If you put a slight crease down the center of each piece, it will straddle the hip easily. The flashing pieces should be slightly narrower than each hip unit so they don't show, yet long enough so that each piece laps the previous one by a good 3 in.

Valleys—It's critical that valleys get done properly, since the roof directs water right to them. Valleys can be open or closed (drawings, above). The open valley can be shingled faster and cleaned more easily. In closed valleys, leaves and pine needles can be a problem, and some people don't recommend them. But I prefer their neat look. Shingles butt up tight in the valley with each course flashed much like Boston hip flashing.

For a closed valley, start with a 36-in. piece of 30-lb. felt laid in the full length of the valley, then add a piece of 12-in. by 12-in. flashing cut diagonally in half. Add the starter course of shingles. Next comes a 12-in. square piece of flashing, and the second course of shingles.

Each valley shingle must get a miter cut along its inner edge where it meets the other valley shingle. It is often easier to cut these on the ground, and do your final fitting on the roof with a hatchet or block plane. Use the bedsheets that you saved. If you are cutting ahead of yourself on the ground, don't forget to cut both right and left-handed shingles. The flashing pieces can also be cut ahead. Remember when you nail both shingles and flashing to use only one corner nail as far away from the center of the valley as you can on each side.

For an open valley, also lay in a 36-in. wide sheet of 30-lb. felt; it's a good bed for the metal that follows. On pitches under 12-in-12, use 20-in. wide sheet metal, which should be nailed at the extreme edges with fasteners that are compatible with the flashing (see *FHB* #9, p. 50). Tin, lead, zinc and galvanized steel are all right for valley metal, but I like to use copper. There have been a few cases of copper flashings and cedar shingles reacting chemically to produce a premature corrosion of the copper. I have never seen this, nor has anyone I know. Just to be safe, you can use a cant strip to minimize the contact between the shingles and the copper (drawing, above right). Whatever flashing you use, it will be improved by a crimp in the center.

I often begin in the valley by nailing the first few shingles on five or six courses high before I carry the courses across the roof. This stacking allows me to set the valley shingles to a chalkline without being restricted by having to fit them to the shingles in the field. Just as with closed valleys, it is important to nail as far from the valley center as possible. □

Bob Syvanen is consulting editor to Fine Homebuilding, *and a builder on Cape Cod.*

Busting Shingles

Tools and techniques for making and applying oak roof shingles

by Drew Langsner

I began making shingles because they seemed to be the right roof for the hewn-log house I was building. We live in southern Appalachia, and I like the idea of using native materials when possible, but the main reason for making shingles is that they make a beautiful roof.

From the outset, I knew that wood shingling is not the most practical roofing available. First, there is the risk of fire, so you have to be very careful when installing the flue for your fireplace or woodstove. Second, making shingles is time-consuming; if you don't like repetitive work, then forget about it.

A shingle roof is not cheap either. It could be—if you didn't put a value on time, and if you had free access to good shingle logs. I had to buy mine. But framing our roof was far less expensive than framing for conventional roofing. For rafters, I used slender poplar saplings. For shingle nailers I used locally sawn boards. The house has a cold attic (insulation is in the ceiling of the upstairs bedrooms), which is best for shingles, since they last longer with good air circulation. So there was no cost for plywood decking or builder's felt.

The question about shingles that I'm most often asked is "How long will they last?" I really don't know. The answer depends on a group of factors—variety and quality of wood, how the shingles were applied, and exposure to weather. Last year my friend Daniel O'Hagan replaced a red oak shingle roof that he had nailed down in 1958—that's 26 years. White oak shingles should last considerably longer, maybe twice as long.

Our shingles are red oak, mainly because of the scarcity of a shingle grade of white oak. A shingle log must be the highest quality—clear of knots, straight grain, no twist. Most important, the wood must split cleanly, into predictable divisions as thin as 5/16 in. Shingle-quality red oaks are not easy to find, but appropriate white oaks are even scarcer. Veneer-grade white oak logs sell for about twice as much as red oak logs of comparable quality.

The surface area of our roof is ten squares. The four red oak logs that I bought cost about $350. The extra cost of white oak would have been a good investment. When you consider the time put into splitting and shingling, white oak is a better value because it lasts longer—if you can find the logs, and put the cash up front.

At $35 per square, the cost of red-oak shingle logs (just the raw material) is comparable to buying asphalt shingles and economy-grade metal roofing. But asphalt shingles and metal roofing are sold ready to go, and application is fast.

Here's a rough breakdown of the time it takes to make oak shingles, based on my experience (with three roofing projects) and those of several other builders I've talked to. The shingles I make average just over 4 in. in width, and they are applied with 6 in. to the weather (photo below). One square requires about 500 shingles. On a good day I can make 200 shingles.

Splitting the 5,000 shingles for our ten-square roof required 25 days of labor. Applying oak shingles also takes longer than using commercially available cedar shingles. Getting nails started in hard oak requires some care, and the shingles aren't easy to trim either. One square per day is possible if you work at a good clip, and if the roof isn't so steep that you need to work from roofing brackets or a shingler's seat. I applied most of the shingles for our roof by standing on scaffolds in the attic. Application took about twenty days.

Selecting logs—For efficient shingle-making, the logs should be straight and free of knots. Trees should be at least 24 in. in diameter, measured at chest height. At the tip end, logs should be at least 18 in. across. Anything less will result in impractically narrow shingles and excessive waste. If knots are unavoidable, count on waste wood on all sides of the knot.

Log taper should be gradual. Taper will be reflected in the shape of the shingles, requiring a lot of trimming. The cross section should be round. Any unevenness in the log will contain wood that is distorted and unusable.

It's very important to look for twist. Vertical furrows in the bark should run straight up and down. A spiral pattern is a sure sign of twisting grain, and a definite reject log.

Finally, some species of red oak and white oak do not split cleanly. I've heard mountain lore on this subject, but I don't know what's actually true. The live oaks are obviously excluded. Some other oaks, such as swamp white oak, are almost impossible to split.

If you decide to cut your own timber, you should be aware that this is serious logging. Green oak weighs as much as 75 lb. per cu. ft. Most trees that would make shingles will be over 100 ft. tall. When you cut one of these giants you're dealing with tremendous forces. If you're new to this scale work, get some experienced help. Bucking a tree into logs can also be dangerous. And you will need to move the logs.

Shingles will most likely come only from the butt section (the lower 12 ft. to 20 ft.). Above that you'll probably have potential sawlogs, plus a good haul of firewood. Unless your shingling project is very small, you'll be dealing with quite a few trees.

As much as I enjoy cutting timber, I prefer to buy my shingle logs, usually from a sawmill (other sources are private woodlots and log brokers). At a sawmill the logs will be separated by species and quality. Since the logs are already bucked to length, you can get a good look at exactly what's available. Look for checks, interior rot and sap stain (bluish blobs on the end grain that can indicate interior rot, or sometimes metal). Logs cut within the last year should be in good condition and split nicely. Sapwood rots quickly, but it's trimmed off the shingles anyway. Since

Splitting out bolts. **A 20-in. length of oak log is first split into halves (right). Pencil lines every 3½ in. around the perimeter indicate further subdivisions, which are then split open with sledge and wedges (the large wooden wedge on the log is called a glut). Next the log is split into segments that are at least 2½ in. wide at their innermost cross split (below right), and the segments are separated.**

shingles are split radially (across the growth rings) you don't need to worry about deep end checks, which would ruin sawlogs.

When you find a likely log, ask the mill operator to pull it from the stack. Take one more good look at your potential shingle wood. Be aware that you really don't know what you're getting until you actually split out some shingles.

If the log looks good, buck it into blocks (shingle-length sections) on the spot. I usually do this myself, but I sometimes ask someone at the mill for help. These chainsaw cuts will be the end cuts of the shingles, so it's important to make a straight, clean cut with your saw. My shingles are 20 in. long, so I measure 20-in. increments from the tip end of the log, marking with a timber crayon or with V-cuts made with an ax. An odd-length waste block at the butt is preferable for two reasons: butt wood is often wavy-grained, and buttress flare is a nuisance at best.

From blocks to bolts—To break up the log blocks into smaller chunks, called bolts, you'll need a variety of tools. For measurements and marking I use a 1-ft. ruler, a pair of straight-leg dividers (optional), and a water-soluble pencil. You'll also need an iron maul (a splitting maul or a sledgehammer), a fairly sharp ax, three iron wedges, and two large wooden wedges, called gluts. Gluts are used to widen a crack that has been opened by an iron wedge. They make it easy to remove stuck iron wedges, and allow you to chop cross fibers in a crack without risk of damaging your ax blade.

Most oak logs have at least an incipient crack radiating from the pith. You should always halve the block by finishing the original split. Notice the ray lines that radiate from the pith. The rays are elongated cells that split easily. Plan to split the rays whenever possible. Drive one or two iron wedges along the ray line at the end of the natural split. Try always to leave at least 1 in. of the wedge above the log surface. It may be necessary to knock the wedge left and right a few times to get it out. Then drive the glut into the widest part of the crack. Locate the glut at a clear opening. A dry hardwood glut will withstand a lot of pounding from an iron maul, but will self-destruct if it runs into cross fibers in the split. The block should split into halves, with few cross fibers connecting one piece to the other. With a very large block, it could be necessary to drive an extra wedge or glut. If it doesn't split at this point, you may have to reject the log.

Set the dividers at 3½ in., and step off 3½-in. increments at the first ring of heartwood, since all sapwood is unsuitable for shingles and must be rejected. Make a pencil mark at each division. (You could also use a ruler.) Don't worry about an odd-width end segment. Now pencil in

lines toward the pith following the most prominent ray near each division mark. You are trying to make divisions that will leave equal pressure on either side of a split so the wood will split cleanly. Count the marked-off bolts in each block half. We want to split the block into units of four bolts, since these will easily divide into halves and quarters. If we end up with a cluster of three bolts, that's okay too. The equal-pressure rule does not become powerful until you get into the divisions within one 3½-in. wide bolt.

Splitting out the bolts is basically the same as busting the main block. Here are a few tips. Start with the larger splits, at clusters of four bolts each (photo top). Then subdivide each cluster. Drive the first iron wedge just inside the heartwood zone. Drive another iron wedge a few inches in toward the pith. Remove the first wedge as soon as it loosens—otherwise it will drop down into the split, where it could block the use of a glut, or catch the edge of your ax. Drive the glut at the location of the first wedge. The oak should split clean through. To remove the glut, hit its inner side to loosen it. Don't pull

the sections apart, or chop any cross fibers until you've split all the bolts. By leaving all the segments in place, they will support each other until the final splits are complete.

The next step is to split the bolts across the rays roughly parallel to the growth rings (bottom photo). Set the dividers at 2½ in. (or use your ruler). Locate the place on each bolt that is 2½ in. across, and pencil a line straight across the bolt. Since 2½ in. divided by 8 is 5/16 in., this will be the minimum thickness of each shingle at its butt end.

Logs greater than 24 in. in dia. will often yield an inner ring of bonus shingles. Set the dividers at 1¼ in. to locate the innermost growth-ring cross split. The shingles must be at least 3 in. wide to be worth bothering with. Try to make the maximum number of shingles. By taking risks, your skills will improve. Also, many failures can be salvaged by sawing them to the shorter lengths needed for the starter and ridge courses on the roof. When all the splits are finished, separate the block into individual bolts. A good shingle log will virtually fall apart. To get

Riving and dressing shingles. With the bolt held firmly by a brake made from scrap lumber nailed to an old stump, Langsner strikes the froe smartly with a club to begin the split (left). Once the blade has entered the wood, he uses the froe as a lever, dividing the piece in two (above). Shingles are then edge-trimmed with a hatchet, and thicknessed with a drawknife while being held in a shaving horse (bottom). The drawing below shows how to make gluts and froe clubs from hardwood saplings.

Making froe clubs and gluts

Wooden gluts and froe clubs can be made from green, dense hardwood saplings (hickory and oak are good) about 4 in. in dia. at the height of the first major limbs. Gluts and clubs made from green wood must season at least a month before use.

Froe clubs: saw off an 18-in. length, leaving at least one good knot in the club head (the handle end should be clear). Then shape the handle to a comfortable size (1½-in. to 2-in. dia.) with an ax or a drawknife, as shown at left.

Gluts: Saw a length of sapling about 3 in. in dia. and 3 ft. to 5 ft. long. With the pole balanced on the chopping block, shape the end into a wedge shape, as shown below. Saw the glut off the pole about 10 in. long. Make two or three more.

10 in.

8 in.

|← 8 in. →| |← 2 in. →|

#1   #2   #3

some of the inner bolts, it may be necessary to reintroduce wedges from the opposite end of the log.

Splitting out the shingles—Now the real fun begins—riving the bolts into shingles. For this you'll need a froe, a froe club, a shingle brake and a bench-height chopping log. The brake is a holding device that uses counterpressure against two rigid cross members. A forked tree limb spiked across two posts can be used. The fork angle should be narrow, so that different widths of stock can be worked between the legs. My brake (photo far left) was made from scrap 1x4s and 2x4s nailed to an old stump.

The froe is an L-shaped splitting tool. The horizontal blade has a socket at one end to accept the vertical wooden handle. The tool is simple, but good ones (old or new) are not common. You can buy a factory-made froe blade (manufactured by Snow & Neally and sold by Woodcraft Supply, 41 Atlantic Ave., Box 4000, Woburn, Mass. 01888 and other mail-order outfits) and modify it by grinding its V-edge profile to an angle of 30° to 35°, and smoothing the transition between the bevel and the side of the blade. The top edge of the blade, which the froe club will strike, should have its edges eased. Factory blades are about 14 in. long; I find a 10-in. length adequate. You can also have a blade made to your specifications by a welder (or a blacksmith who knows how to do forge welds). Froe clubs can be made from any dense, heavy hardwood (drawing, left). Mine are made from knotty hickory saplings.

Here's the splitting procedure. Balance any 3½-in. wide bolt on the stump. It doesn't matter which end you split from. Place the froe parallel to the rays (crossing the growth rings), along an accurately located, imaginary mid-line. Hold the froe steady as you raise and lower the club. (Beginners often have trouble with involuntary shifting of the froe position just before striking; holding the handle near the blade will help.) Strike the froe with a determined blow (photo far left). The blade should enter the bolt with the first hit. If the froe doesn't penetrate, place it in the exact position of the first strike and try again. (If you relocate the froe, you'll be opening two cracks, causing loss of splitting control, and possible loss of a shingle.) Continue striking until the froe blade has penetrated to its full width.

Now put the club aside, and insert the bolt (with attached froe) into the brake. Pull the froe handle down toward your body. Your bolt should split into neat halves. If it partially splits, slide the froe forward, then pry again. Try to refrain from driving the froe down with the club. The froe should be used with a lever action (photo above right), not as a driven wedge.

Since the sapwood is much less decay resistant than the heartwood, the next step is getting rid of it. If the sapwood is thick and already decaying, I sometimes wedge it off before I separate the split bolts from the round block. Strike the froe along the growth ring dividing sapwood and heartwood. The split will wander toward the bark side of the bolt. This run-out is the effect of uneven side pressure. Reverse the bolt to split the sapwood from the other end. Any sapwood that remains will be axed off later when the shingles are dressed.

Now rive a half-bolt lengthwise (parallel to the ray plane). Riving skill is a matter of practice more than anything else. Don't rush. Observe the split as it opens. If the division begins to be uneven, with the narrower portion toward yourself, rotate the bolt 180° so that pulling the froe levers against the thicker side. The split will move back toward the center.

You now have two quarter-bolts, each about ⅞ in. thick. Proceed to split these again. Good shingle oak easily splits to 7⁄16 in. Accurate positioning of the froe at the midline of the quarter-bolt is critical. Aim for perfect division so you won't have to make corrections during the split. Controlling a split takes practice. It's important to take action at the first indication of run-out. Eventually, you'll be able to feel uneven splitting before you can see it. Also, run-out control is more effective (and subtle) with the final splits.

Trimming and thicknessing—Dressing the shingles consists of edge-trimming with a hatchet and thicknessing with a drawknife. The objective is to make flat shingles, 5⁄16 in. to 7⁄16 in. thick at the butts, and ¼ in. to 5⁄16 in. thick at the tips. I've heard about old-timers who could split 500 shingles a day. This may not be an exaggeration. But I'll bet they did little if any dressing. Depending on riving skill and wood quality, dressing can take as long as splitting.

From my experience, it's worthwhile doing first-rate dressing work. As with masonry, when you apply a course of shingles you are simultaneously preparing the ground for the following row. Application will go faster and be better if your shingles are an even thickness and lie flat.

For edge trimming, you'll need a very sharp broad hatchet, and a log stump to work on. A broad hatchet is beveled to a cutting edge on only one side of the blade. The other side is flat, which means that you can pare straight down the edge of a shingle. It works like a cutting tool, rather than just splitting. Broad hatchets are fairly easy to locate. Plumb (The Cooper Group, 3012 Mason St., Monroe, N. C. 28110) still makes them, and several nice imported versions are offered by mail-order tool companies.

The object of side trimming is to remove any remaining sapwood, and to make the shingle sides parallel. Hold the shingle roughly vertical, with your fingers well out of the way of any ax action. Now place the hatchet at the required location. Lift the shingle and the hatchet simultaneously. Then come down on the stump, using a little extra force with the hatchet hand. If you're not an expert axman, this is considerably safer and more accurate than taking a freehand swing at the shingle.

The drawknife is used to thickness and flatten shingles that don't meet specifications. Most shingles will require some drawknife work, if only to give the tip end some taper. Any sharp drawknife will work. But since we're going to be dressing thousands of shingles, I recommend getting one that's suited to the purpose. I use a Peugeot drawknife. Made by the car company, it's sold as a French drawknife by mail-order houses. The blade is slightly convex, with the

bevel on the upper side, making this an excellent tool for removing wide, thin shavings from a flat surface. In addition to the drawknife, you'll also need a shaving horse to hold the shingles while you work (photo facing page, bottom).

When you pick up a shingle, the first thing to check for is wind (or twist). Look at the shingle from one end. If it's flat, the near and far edges will look parallel. If the shingle is at all thick, wind can be eliminated by selectively drawknifing wood from the high areas. If removing the wind requires thinning past the minimum thickness, you have a reject shingle; it may try to twist again if it's nailed to the roof.

When drawknifing, aim to remove wide, thin shavings. By skewing the drawknife at an angle to the shingle, you get a narrower effective cutting angle, and a better slicing cut. If the shingles tend to bend from drawknife pressure, insert a 1x3 extension stick between the shingle and the lower jaw of the shaving horse. You may want to screw the extension stick in place, so it doesn't need to be adjusted constantly. If you come across a really thick shingle, set it aside. When you have a small pile of these, froe off the extra thickness by using controlled run-out.

Tie and dry—After dressing comes bundling, tying the shingles into transportable units with twine. Gather the shingles into groups of 20, and put them into a fully opened woodworker's bench vise. Arrange the shingles so that the widest ones are in the center, grading to the narrowest shingles at the sides of the bundle. Rotate or reverse shingles so that they fit together in a compact mass. Close the jaws just short of busting any shingles. (This is a dare.) Then tie a length of twine around each end of the bundle.

I think shingles should be air dried, not green, when they're applied to the roof. This is contrary to local folk wisdom, which recommends nailing shingles green, and when the moon is waning. ("If you nail 'em when the moon is waxin' they'll quarrel up on you.") In my opinion, shingles curl up when they're nailed green because the top side dries out faster than their underside. When shingles are nailed dry, there isn't a large moisture differential between the exposed side and the underside. Nailing dry shingles takes more care, but in my experience only one in several hundred will split. Given fair weather, oak shingles air-dry to equilibrium moisture content in about ten days. Set the bundles in a well-ventilated shelter to dry them.

Bundles of shingles, dipped in preservative, are set out to dry.

Preservatives—After air-drying, the shingles can be treated with preservative. Over the past five years I've tried various treatments, but the circumstances changed from one case to another, so the results can't be compared. The commercial products I've used are Clear #20 Cuprinol (Darworth Inc., Box K, Tower Lane, Avon, Conn. 06001) and Water Beader (Pratt & Lambert, Box 22, Buffalo, N. Y. 14240). Both cost about $65 for five gallons. I use the treatments as a dip, before the shingles are applied. For 500 shingles (about one square), you'll need about three gallons. Clear Cuprinol and Water Beader contain fungicides. Clear Cuprinol has zinc naphthenate; Water Beader has "549M" (3-iodo-2 propynyl butyl carbamate). With either of these, you're supposed to re-treat after five years, then every ten years.

For dipping I use a 5-gal. plastic bucket, filled about two-thirds full. A bundle of 20 shingles is immersed for about 30 seconds, then turned end-for-end for another half minute. While one bundle is soaking, I wedge an extra shingle or two into the next bundle to compensate for shrinkage during seasoning. Bundles are drained in a second bucket, then set aside to dry, which takes three or four days. The bundled shingles are now ready to haul up on the roof.

On the roof, at last—Oak shingles are applied like commercial cedar shingles. Use hot-dipped galvanized nails, which won't react with acids in the oak and rust away. Figure on 50 lb. of 6d nails for every five roofing squares. When a nail is started, dry oak shingles tend to vibrate. Press down firmly on the shingle with the heel of your hand while holding the nail in place. □

Drew Langsner is a freelance writer and director of a crafts school, Country Workshops. He lives in Marshall, N. C.

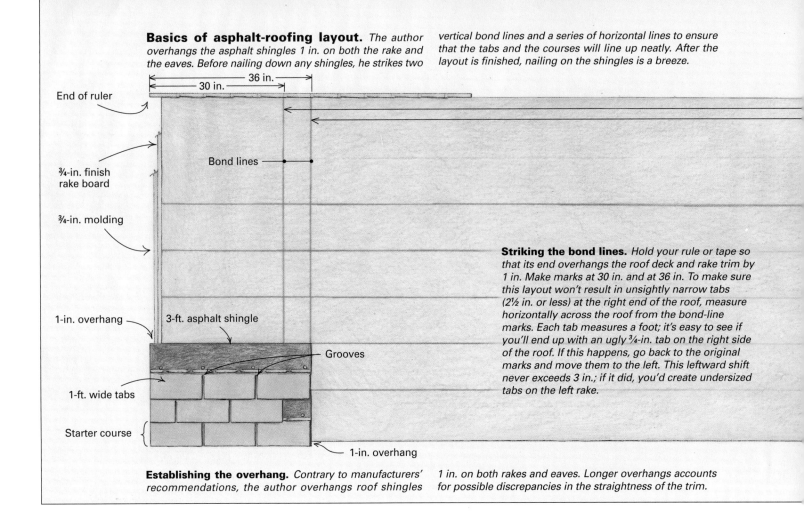

Basics of asphalt-roofing layout. *The author overhangs the asphalt shingles 1 in. on both the rake and the eaves. Before nailing down any shingles, he strikes two vertical bond lines and a series of horizontal lines to ensure that the tabs and the courses will line up neatly. After the layout is finished, nailing on the shingles is a breeze.*

End of ruler

¾-in. finish rake board

¾-in. molding

36 in.

30 in.

Bond lines

Striking the bond lines. *Hold your rule or tape so that its end overhangs the roof deck and rake trim by 1 in. Make marks at 30 in. and at 36 in. To make sure this layout won't result in unsightly narrow tabs (2½ in. or less) at the right end of the roof, measure horizontally across the roof from the bond-line marks. Each tab measures a foot; it's easy to see if you'll end up with an ugly ¾-in. tab on the right side of the roof. If this happens, go back to the original marks and move them to the left. This leftward shift never exceeds 3 in.; if it did, you'd create undersized tabs on the left rake.*

1-in. overhang

3-ft. asphalt shingle

Grooves

1-ft. wide tabs

Starter course

1-in. overhang

Establishing the overhang. *Contrary to manufacturers' recommendations, the author overhangs roof shingles 1 in. on both rakes and eaves. Longer overhangs accounts for possible discrepancies in the straightness of the trim.*

Laying Out Three-Tab Shingles

Spend a little time measuring and striking lines, and the rest is fairly easy

by John Carroll

Many roofers take pride in the fact that they can shingle a house without the benefit of measured lines. It can't be denied that such people install leak-proof roofs that look pretty good from the ground. Unfortunately, their eyeballed roofs often have wavy, inconsistent courses; and when viewed from atop the house, they look simply unprofessional. When I finish a roof, I enjoy looking at straight courses, and I don't begrudge myself the half-hour or so it took to measure and strike lines. More than that, I'm convinced that I recover the time invested in laying out the roof as I nail down shingle after shingle without worrying about the courses getting wavy or crooked.

For the sake of simplicity, I'll limit my discussion to the ubiquitous three-tab asphalt roof shingle, scorned by aesthetes but found on houses from the Carolinas to California.

Running straight courses. The author aligns shingles using a gauged roofing hammer and vertical bond lines.

On a rectangular roof without dormers, valleys or other obstructions, there are three basic layout steps: establishing the overhang, striking the bond lines and striking the horizontal lines.

Establishing the shingle overhang—Before shingling a roof, it is essential to know how far the shingles will overhang the bottom (or eaves) and sides (or rakes) of the roof deck. Ideally, all trim has been installed along the roof edges, and if used, metal drip edge is also in place. In these cases I leave a 1-in. overhang along the eaves and the rakes of the roof (drawing above). Most shingle manufacturers recommend a ¼-in. to ⅜-in. overhang, presumably to reduce the chance of the wind snagging the edge of the roof.

Unfortunately, eaves and rakes (especially those on older houses) often diverge more than

Photo this page: Jefferson Kolle

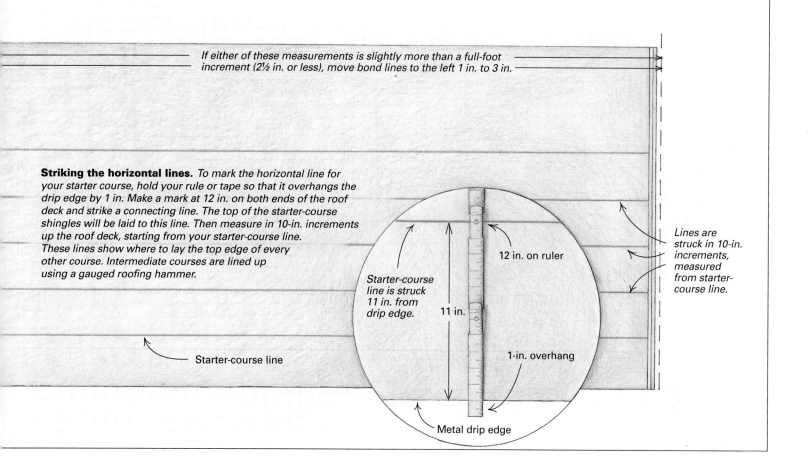

If either of these measurements is slightly more than a full-foot increment (2½ in. or less), move bond lines to the left 1 in. to 3 in.

Striking the horizontal lines. *To mark the horizontal line for your starter course, hold your rule or tape so that it overhangs the drip edge by 1 in. Make a mark at 12 in. on both ends of the roof deck and strike a connecting line. The top of the starter-course shingles will be laid to this line. Then measure in 10-in. increments up the roof deck, starting from your starter-course line. These lines show where to lay the top edge of every other course. Intermediate courses are lined up using a gauged roofing hammer.*

Starter-course line is struck 11 in. from drip edge.

12 in. on ruler

11 in.

1-in. overhang

Metal drip edge

Lines are struck in 10-in. increments, measured from starter-course line.

Starter-course line

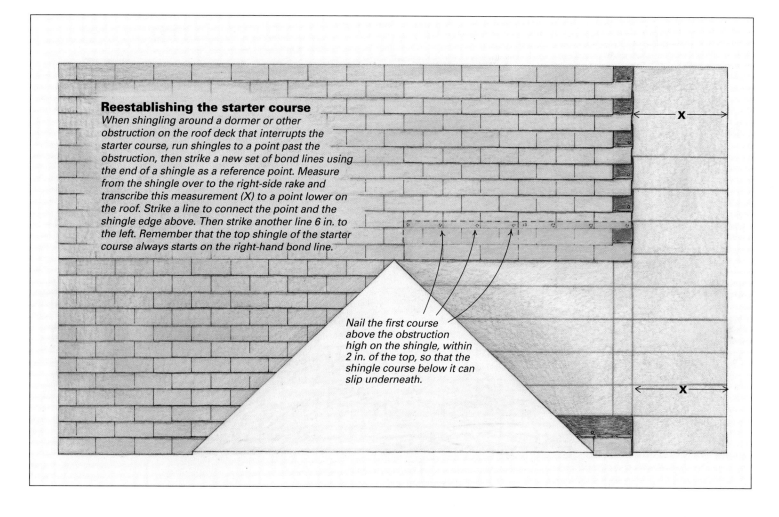

Reestablishing the starter course
When shingling around a dormer or other obstruction on the roof deck that interrupts the starter course, run shingles to a point past the obstruction, then strike a new set of bond lines using the end of a shingle as a reference point. Measure from the shingle over to the right-side rake and transcribe this measurement (X) to a point lower on the roof. Strike a line to connect the point and the shingle edge above. Then strike another line 6 in. to the left. Remember that the top shingle of the starter course always starts on the right-hand bond line.

X

X

Nail the first course above the obstruction high on the shingle, within 2 in. of the top, so that the shingle course below it can slip underneath.

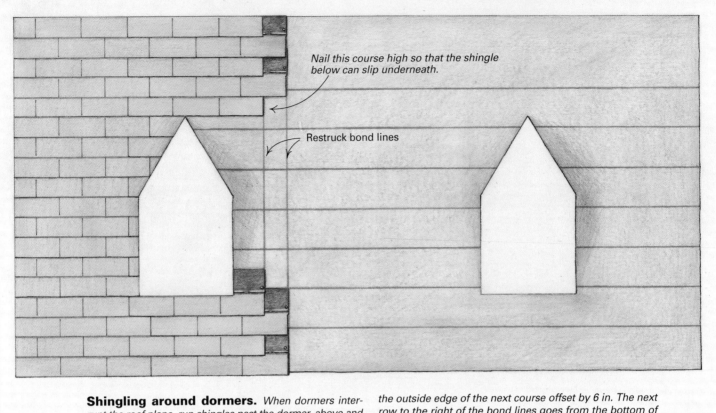

Nail this course high so that the shingle below can slip underneath.

Restruck bond lines

Shingling around dormers. *When dormers interrupt the roof plane, run shingles past the dormer, above and below. Strike bond lines between the top and bottom sections, first along the outside edge of one course, then along the outside edge of the next course offset by 6 in. The next row to the right of the bond lines goes from the bottom of the roof to the peak, leaving an unroofed area to the right of the dormer, which can be filled now or later.*

⅜ in. from a straight line. To compensate for irregularities in the straightness of rake boards and fascia boards, I've found that a 1-in. overhang allows me to work proficiently and provides for a straight, secure roof. I've never had a problem with shingles blowing off.

If I install the shingles before the roof trim is complete, I need to know how far the trim pieces and the drip edge will extend the roof deck, and I allow for that extension. For example, if a 1x6 rake board and a piece of molding, totaling 1½ in., will be added to the existing sheathing, I know I should let the shingles overhang the sheathing by 2½ in. to get a final overhang of 1 in. Because shingles are easy to cut, it is better to err on the side of too much as opposed to too little overhang. If need be, I can go back later and trim the overhanging shingles.

Three-tab asphalt shingles are 3 ft. long. They have three 1-ft. tabs with grooves cut between them in the part of the shingle that is exposed to the weather. The grooves both break up the otherwise solid appearance of the shingle (possibly making them look more like wood shingles or slate shingles) and provide a channel for water to run off the roof. It is important that the grooves of every other course line up over one another and that the grooves of the course in between fall in the middle of the tab of the shingle above and below.

Like most right-handed roofers, I usually start shingling on the left side of the roof. This enables me to work from left to right, positioning shingles with my left hand and nailing them off with my right hand.

Striking the bond lines—To keep the grooves straight and the shingles properly bonded or centered over the tabs just below, the shingles are laid to follow two vertical chalklines, called bond lines, struck near the left rake of the roof (drawing p. 30). Bond lines are always struck 6 in. apart—half the width of a tab; this aligns the 1-ft. tabs of alternate vertical shingle rows.

If I need to leave, say, a 2½-in. rake overhang, I extend my ruler exactly 2½ in. past the roof deck and make marks at 30 in. and at 36 in. This is a preliminary measurement. To make sure this layout won't result in unsightly narrow tabs (2½ in. or less) at the opposite edge of the roof, I measure across the roof from these marks. Each foot represents a tab, and it's easy to see if I'll end up with an ugly ¾-in. tab on the right side of the roof. If this is the case, I go back to the original marks and move them to the left. This leftward shift never exceeds 3 in.; if it did I would be creating undersized tabs on the left rake.

When I'm satisfied that I won't end up with little tabs at either end of the roof, I make identical measurements at the top and the bottom of the roof along the left-hand side. Then I strike a vertical, parallel bond line at both the 30-in. and 36-in. marks.

When I'm ready to install the shingles, I begin each horizontal course on a bond line, alternating between the two bond lines. But before I can

nail on any shingles, I also have to measure and strike horizontal lines

Striking the horizontal lines—Standard shingles are 1 ft. high. To lay out the first course, called the starter course, I need to know the overhang at the roof eaves. If all of the trim and drip edge has been installed, I hold my folding rule so that it extends 1 in. past the drip edge, and I make a mark at 1 ft. I make the same mark at the other end of the roof and then strike a chalkline across the roof deck (top drawing, p. 31).

Shingle exposure is the height of the shingle that will be exposed to the weather. In most cases, the exposure of three-tab asphalt shingles is 5 in. Shingles are 1 ft. high, so each successive shingle will overlap the one below it by 7 in.

It's not necessary to strike lines every 5 in.; in fact, I always strike lines in increments of 10 in. There's a reason for this, which I'll explain shortly. When I mark my horizontal lines, I place the end of my rule at the starter-course line, and then I make marks every 10 in. If I'm working alone, I often strike lines in increments of 20 in. or 30 in. The most important thing to remember is that all lines are measured off the starter-course line rather than off the drip edge.

Running the shingles—After striking lines, I start nailing shingles where the bond lines intersect the starter-course line. The starter course is always nailed on the roof upside down. The next row of shingles is nailed right-side up, directly on

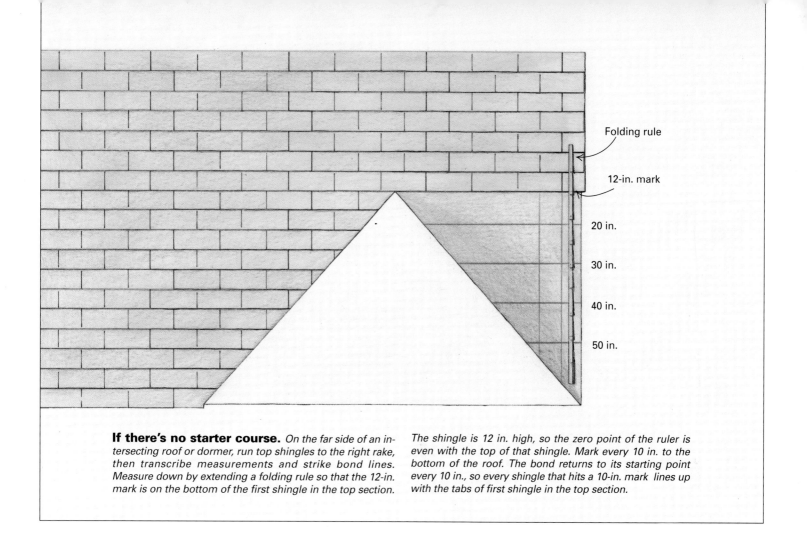

If there's no starter course. *On the far side of an intersecting roof or dormer, run top shingles to the right rake, then transcribe measurements and strike bond lines. Measure down by extending a folding rule so that the 12-in. mark is on the bottom of the first shingle in the top section.*

The shingle is 12 in. high, so the zero point of the ruler is even with the top of that shingle. Mark every 10 in. to the bottom of the roof. The bond returns to its starting point every 10 in., so every shingle that hits a 10-in. mark lines up with the tabs of first shingle in the top section.

Folding rule

12-in. mark

20 in.

30 in.

40 in.

50 in.

top of the starter course. The reason for this is to cover the metal drip edge that would otherwise be exposed to the weather by the grooves in the right-side up second course.

I always begin the upside-down starter course on the left-hand bond line. The next course goes directly on top of the first and begins on the right-hand bond line. Because the lines above the starter course are marked in increments of 10 in., every other shingle hits a horizontal line, and every shingle that hits a horizontal line also hits a right-hand bond line (including the exposed starter). I follow this routine religiously because the consistency is very useful on complex roofs, as we shall see.

Horizontal, diagonal or vertical shingling?— A neat, professional roof can be installed by running shingles horizontally, diagonally or straight up the roof. Running each course horizontally across the roof is the simplest method and is usually preferred by amateurs. Running the shingles diagonally across the roof so that they look sort of like a staircase is often recommended by shingle manufacturers because of the possibility that the shingle color might vary from bundle to bundle. The thought is that the variegations will be less noticeable if the different colors are run diagonally rather than straight up or straight across a roof.

Like many roofers, however, I prefer to run vertical rows straight up the roof. I do this for two reasons. First of all, I find it less strenuous be-

cause it does not require as much reaching and moving about. Secondly, on hot days I find it to be more comfortable because I'm sitting or kneeling on shingles I've just laid. These are a lot cooler than those that have had a chance to soak up the sun. I've never had a complaint about the blend of colors on any of the roofs I've installed. I have noticed, though, that an off-color bundle looks equally bad whether it runs straight up the roof or diagonally.

For those who choose to run shingles vertically, here is one caution: You have to leave the far right-hand nail out of every other course (the one that hits the right-hand bond line). This allows the shingles in the next row to slip into place. I always use four nails to the shingle in the recommended pattern. To do this I have to lift the tab of every other shingle in the preceding row.

Using a gauged hammer—As mentioned previously, I often strike horizontal lines every 20 in. or 30 in. To keep in-between courses straight, I use an Estwing gauged roofing hammer (Estwing Mfg. Co., 2647 8th St., Rockford, Ill. 61109-1190; 815-397-9521). This hatchetlike hammer has a steel knob bolted through its blade exactly 5 in. from the face of the hammer head (photo p. 30). After following the struck horizontal line with one shingle, I line up the next three courses (if I'm using 20-in. increments) with my hammer. The steel knob, or gauge, hooks onto the bottom of the shingle in the previous row, and the bottom of the next shingle sits on the hammer head.

Laying out complicated roofs—So far I've limited this discussion to a straight, rectangular section of roof. Roof planes come in a variety of shapes and sizes, however, and they are apt to be intersected by chimneys, dormers and adjoining roofs. Shingling around these obstructions complicates the job, but by adhering to a consistent 10-in. layout scheme and using a few simple techniques, it's easy to keep the courses straight and correctly bonded.

To go around a pair of dormers (drawing facing page), I lay out the bond lines and the horizontal courses as previously described. Some of the horizontal lines are interrupted by the dormers and have to be measured and marked separately on each side of the dormers. When I start roofing, I run a row of shingles all the way up the left rake and work toward the right until I come to the left side of the first dormer. I continue to shingle the area below the dormer until I'm past the dormer. At this point I move back to the left side of the dormer, cut and fit shingles along the dormer wall, install flashing and weave the first valley created by the dormer's roof.

There is now a short row of shingles running from the top of the valley to the ridge of the main roof. I carry these courses to the right until they line up with the courses below. To permit the courses that will be installed below these shingles to slide into place, I nail the first course high on the shingle, within 2 in. of the top edge. I strike bond lines between the top and bottom sections, holding the string first along the outside edge of

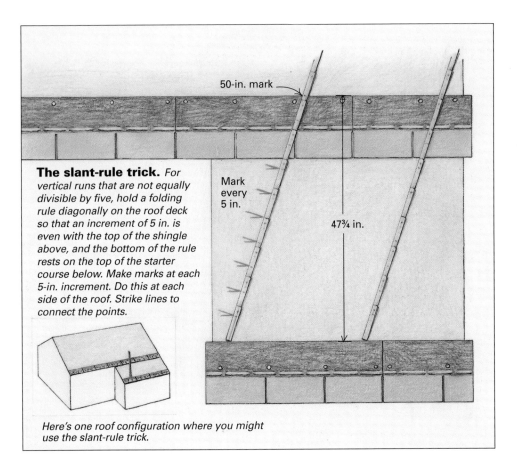

The slant-rule trick. *For vertical runs that are not equally divisible by five, hold a folding rule diagonally on the roof deck so that an increment of 5 in. is even with the top of the shingle above, and the bottom of the rule rests on the top of the starter course below. Make marks at each 5-in. increment. Do this at each side of the roof. Strike lines to connect the points.*

50-in. mark

Mark every 5 in.

47¾ in.

Here's one roof configuration where you might use the slant-rule trick.

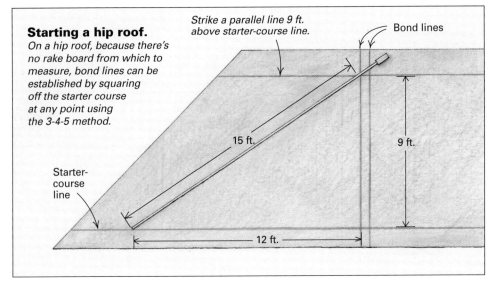

Starting a hip roof.
On a hip roof, because there's no rake board from which to measure, bond lines can be established by squaring off the starter course at any point using the 3-4-5 method.

Strike a parallel line 9 ft. above starter-course line.

Bond lines

15 ft.

9 ft.

Starter-course line

12 ft.

zontal line also hits a right bond line. I put the inverted shingle of the starter course on the left bond line and cover it with the exposed starter course on the right bond line. As I run up the bond lines, I notice that every shingle that hits a horizontal line also hits a right bond line. I know the bond will work out perfectly.

When there's no starter course—Sometimes there's no starter course on the far side of an intersecting roof or dormer (drawing p. 33). If so, after I run the top section of shingles over to the rake, I measure and strike bonds in the usual manner. Let's say that I have not struck any horizontal lines in the triangular section created by the intersecting roof. How would I measure down, and what bond line would I start on? To measure down, I extend my folding rule and lay it on the roof deck so that the 12-in. mark is on the bottom of the first shingle in the top section. The shingle is 12-in. high, so this puts the zero point of the rule even with the top of that shingle. Now I mark at every multiple of 10, i. e., 20 in., 30 in., etc., until I get to the bottom of the triangular section of roof. Because the bond returns to its starting point every 10 in. (or every other course), I know that the grooves of every course that hits a 10-in. multiple will line up with the grooves of the first shingle in the top section. I make sure it does.

Starting a hip roof—Sometimes it's not practical to start roofs on the left side. Hip roofs or roofs with obstructions on the left side should be started toward the center of the roof deck. On a hip roof you can't measure from the rake to establish the bond lines. They must be squared up at some point along the starter course using the 3-4-5 method (bottom drawing, left). After striking the starter-course line, I mark a point, measure 12 ft. along the line from that point and make another mark. I strike a parallel line 9 ft. above the starter course, pull a tape measure diagonally from the 12-ft. mark on the starter course until the 15-ft. mark on the tape intersects the upper chalkline, and I make a mark. Stretching a chalkline from my first mark on the starter course to the mark above to the ridge of the roof, I strike my first bond line. I strike a second line 6 in. to the right. Then I can run all of my shingles from left to right, then come back and fill in the hip.

The slant-rule trick—Occasionally it's necessary to fit shingle courses into a space that's not divisible by 5 in. (top drawing, left). If the run of a section of roof from starter course to intersecting roof is, say, 47¾ in., some roofers might run nine courses at 5 in. and the last course at 2¾ in. A better way to set up the courses is to divide the 47¾ in. into 10 equal courses. Nail on your first two starter courses—inverted and right-side up. Then put your tape on the starter-course line and run the tape diagonally across the roof until you come to a 5-in. increment—in the example above, 50 in. Make a mark at each 5-in. increment. Do this on both sides of the roof and strike lines between each mark. ☐

John Carroll is a builder in Durham, N. C.

one course, then along the outside edge of the next course that is offset by 6 in. The next row of shingles goes from the bottom of the roof to the ridge, leaving an unroofed area to the right of the dormer. This area can be filled in now or later, according to the temperament of the roofer. I like to complete this section as I go along.

This process is repeated around the second dormer; I shingle past the dormer at the top and the bottom, strike bonds lines through and fill in.

Reestablishing the starter course—When large dormers or intersecting roofs interrupt the bottom section of a roof plane, it is impossible to strike through (bottom drawing, p. 31). So after running shingles across the top of the roof, nail-

ing the first course high until I've cleared the entire obstruction, I measure the distance from the end of one of the right bond shingles to the right edge of the roof deck. I transcribe this measurement to the bottom of the roof, make another mark 6 in. to the left and strike bond lines. Then I measure and strike my 10-in. horizontal lines.

I'm now ready to run shingles from the bottom of the roof up the bond lines. But the question is, on which bond line do I start? If I pick the wrong one, I'll end up with adjoining courses where all the tabs line up rather than being offset by 6 in.—a roofing abomination. Fortunately, I've struck lines every 10 in., and I've started, as I always do, with the exposed starter shingle on the right bond line. I know every shingle that hits a hori-

Built-Up Cedar Roofing

Common shingles mimick a traditional reed thatch

by Steve Dunleavy

About eight years ago my company was hired to put a roof on a custom home in the Sierra Nevada mountains overlooking Lake Tahoe. The house was a massive Tudor mansion with a steep roof, incorporating turrets, dormers, hips and valleys. The builder had spent a lot of time researching European architecture and wanted the house to look authentic, including the roof, which was to be reed thatching.

True thatch roofs are practically non-existent in North America. Importing material and labor, primarily from England, has proven to be prohibitively expensive—most companies who have tried it have met with failure. Building departments and insurance companies unfamiliar with what amounts to a Middle Ages technology have been hesitant to approve or underwrite what my dictionary calls "a roof covering made of straw, reeds, leaves, rushes, etc." Also, I've seen thatch roofs play host to every sort of insect, bird, rodent and rot imaginable. For all these reasons, the reed-thatch plan was scuttled and I was asked to simulate, using cedar shingles, the droopy, layered look of authentic thatch roofing. Beginning with that first house and refining and borrowing elements from many styles of shingling, we devised a shingle roof that we call "built-up cedar thatching." My company has now installed more than two dozen of them, and they've proven to be both durable and dazzling to the eye, capturing the velvety look of traditional reed thatch (top photo).

Division of labor—From my standpoint, applying built-up cedar thatching seems more like a military campaign than a roofing job. Everything involved is on a large scale: the intense labor; the massive amounts of material stored and then moved higher and higher up steep inclines; the wear and tear on men, tools and equipment. It's no surprise that labor is the biggest expense with these roofs. Labor costs must be budgeted carefully based on the job, the climate and the topography of the roof itself.

I break my crews into two categories: nailers and laborers (or grunts and sub-grunts, as

Though it resembles a traditional reed thatch, a built-up shingle roof consists of cedar shingles split to a maximum width of 5 in. and laid in random patterns with exposures of 3 in. or less. In the top photo the shingles wrap around radiused rakes and eaves. At hips (photo above), shingles are tapered with a utility knife and laid like pie slices. The angle of the taper depends on the roof pitch. A layer of Ice & Water Shield beneath the roofing felt at both hips and valleys adds protection against the weather.

they've graciously named each other). The nailers are usually journeymen roofers who are familiar with built-up shingle thatching (although experienced laborers occasionally nail, too). Nailers are also responsible for installing weatherproof membranes, roofing felt and metal flashings.

The laborers' job is to keep the nailers supplied with shingles, which is more difficult than it sounds. A built-up shingle thatch requires two to three times the amount of material needed on a standard shingle roof, and every shingle has to be hauled up to the roof via winch or ladder as the job progresses (there's just too much material for a rooftop delivery). Also, laborers are responsible for splitting each shingle to a width of 5 in. or less, which is critical to the appearance of these roofs.

If the roof framing is curved at the eaves, it's usually necessary to steam-bend shingles prior to nailing them up. This is a full-time job for one man (more on that later).

Everyday materials—While a built-up shingle thatch might appear exotic, the materials we use to assemble it are not. We prefer 16-in. long, #1-grade Western red cedar shingles, which are 100% edge grain and clear heartwood. We've tried using 18-in. long shingles, but found them to be harder to split and steam-bend, as well as a little too thick to lie properly. We've had some success with #2-grade shingles, which are clear and knot-free for a minimum 10 in. from the butts. They cost $10 to $20 less per square than #1-grade shingles, so they're occasionally specified by clients to save money. These shingles don't affect the appearance of a roof, but they can be difficult to split and bend because of knots, leading to increased waste and more work. Generally, I figure that a simple cedar thatch roof will be at least three times the cost of a standard roof; complicated roofs run even more. Labor is easily 50% of the job.

Because of ice-dam problems in snow country, we're required by code to install an elastic-sheet membrane over the first nine feet up from the eaves. We use Ice & Water Shield (W. R. Grace and Co., 62 Whittemore Ave.,

Cambridge, Mass. 02140), which comes in 3-ft. wide rolls that cover 225 sq. feet. The product consists of rubberized asphalt with a cross-laminated polyethylene film on one side and a pressure-sensitive adhesive backed by removable kraft paper on the other. The sheeting is installed polyethylene side up, forming an impermeable membrane that stays watertight even when nails or staples are driven through it. Ice & Water Shield has an embossed surface, so it's not slippery to walk on, an endearing feature to roofers.

In addition to using the membrane along eaves, we use it as an extra precaution under the flashing at roof-wall intersections, valleys, chimney saddles and other potential trouble spots. At about $50 per square it's expensive, but well worth its price for the protection it offers. A warning, though: Ice & Water Shield is the stickiest flypaper ever invented. On a steep roof it's a two man operation to put down, and we don't mess with it in any sort of wind.

We are also required to install tar paper over the membrane. We use 30-lb. felt affixed to the deck with barbed roofing nails. The felt is substantial enough to provide temporary weatherproofing and serves as a durable substrate for a built-up thatch.

For flashings we prefer copper, although many people ask for galvanized sheet metal to save money. We fabricate and install most of our own flashings, but if a peculiar bend or solder job is giving us trouble, we call in the experts. The quality of the flashing material and workmanship is critical because these roofs, if installed properly, should last a century or more. There are times when nothing else will work except a tube of gun-grade urethane caulk, so we always have some available in case the need arises.

Tooling up—We prefer to use pneumatic staplers to apply shingles, as does most of the roofing industry. Indeed, roofs like this, typically requiring a quarter million or more fasteners, would be astronomically expensive and virtually impossible to apply if they had to be hand nailed. Our weapon of choice is the Senco MII pneumatic staple gun (Senco Products Inc., 8485 Broadwell Rd., Cincinnati, Ohio 45244), loaded with 2-in. staples. These tools are almost indestructible on the outside (every one I own has cartwheeled off more roofs than my esteemed employees care to admit). When the insides go, virtually everything can be replaced using inexpensive repair kits. I've rebuilt these guns on the tailgate of my truck at the job site. Staples can be bought almost anywhere, either generic or brand name. We use Senco brand staples because the tolerances seem a little tighter and they don't seem to jam or break the tools as often as other staples do. Pneumatic tools are percussive and loud, so we wear disposable foam earplugs, which can dramatically reduce fatigue.

We use a variety of saws in our thatch work. For on-the-roof cutting of shingles we prefer small circular saws, such as the Makita 4200N. Detail work around skylights and chimneys re-

Shingles for radiused eaves are steamed and bent to the proper curvature before installation. The author's steamer consists of a truck utility box modified to fit over a 55-gal. drum of boiling water (top photo). Once shingles are saturated with moisture and become pliable, they're transferred to the bending jig (middle photo). When the jig's lid is closed, a pressure bar in the lid forces the shingles to bend over a platen to the proper radius (bottom photo). Typically, about 10% of the shingles break while bending.

quires plenty of cuts, and these lightweight trim saws are more comfortable and safer to use than are the larger 7¼-in. circular saws (for more on trim saws, see *FHB* #48, pp. 40-43). We use a four-tooth carbide blade, which doesn't make a really precise cut but will last the length of the job (we're not, after all, building a piano here). This combination of high-rpm saw and four-tooth blade is effective but potentially dangerous because of the size and velocity of the waste it expels (the waste is more akin to chain-saw chips than sawdust). Eye protection is a must.

Cutting through several layers of shingles, such as at the ridge of a roof, requires more power than a trim saw offers, so we switch to

a standard 7¼-in. worm-drive saw. Weaving a built-up thatch over hips, curved rakes and other obstacles can require thousands of shingles to be cut at predetermined angles. We accomplish this with a radial-arm saw set up on the ground, and haul tapered shingles up on the roof as we need them.

Bending shingles—The steam boxes and bending jigs we use change from job to job, depending on what's needed. Our steamer consists of an old truck utility box with a metal flange welded on the bottom that fits over a 55-gal. drum full of boiling water (top photo). Water is heated by a propane weed burner at the base of the drum, and a large opening in the bottom of the box allows steam to enter. We load and unload material through the hinged door; a 2x4 rack with a row of nails front and back holds the shingles on edge. A second rack can be stacked on top, allowing the box to hold twice as many shingles. We rotate our stock, replacing hot, pliable shingles with fresh ones. The idea is to be as efficient as possible—we don't leave the steambox door open too long, and we have the bending jig open and ready for shingles when they come out of the steamer. Shingles are usually steamed for 20 to 30 minutes. There's some springback, but not much.

Our bending jig comes from the same make-it-on-the-spot family as the steamer. It can be made out of scrap lumber in a few hours (middle photo). The jig is simply a hinged frame made of 2xs and plywood, with a strategically placed 2x2 bending platen in the base of the frame and a 2x4 pressure bar in the lid. Shingles are loaded into the base butt-end down, with the butts constrained by a 1x4 crossbar. When the lid is closed, the pressure bar forces the shingles into a curvature dictated by the relative positions of the pressure bar and bending platen (bottom photo). We usually have to fine-tune the jig to achieve a particular bending radius, and we figure on breaking about 10% of the shingles.

The jig is screwed together with "grabbers" (similar to drywall screws), both for strength and so it can be adjusted. We intentionally build our jigs small so that we can transfer shingles quickly from the steamer to the bending jig before they have a chance to cool off.

Thatching guidelines—The most important factors for achieving the look of a thatched roof are shingle size, exposure and pattern. A typical non-thatch shingle roof is composed of straight 5-in. courses made up of shingles 4 in. to 14 in. wide, and measures about ⅞ in. thick. A built-up thatch roof, on the other hand, is composed of smaller shingles (5 in. wide or less) laid in a random pattern with exposures ranging from ½ in. to 3 in. The result is a roof surface that's about 3 in. thick.

Breaking shingles to the proper width is more complicated than it sounds. We break them by hand because it's quicker and safer than swinging a hatchet thousands of times.

By grasping a shingle on its side edges with thumbs in the middle and quickly snapping the wrists, we usually get a clean break. Moist shingles can bend almost in half before breaking and require more effort, as do thick or dry shingles. Beginners tend to get sore hands quickly and usually try to break material over their knee caps, then get sore knees and go back to the first method. Sometimes the grain isn't straight, resulting in an angled break. These pieces are set aside to be trimmed later with a knife. Breaking wide material into smaller pieces goes against every good roofer's natural instincts, but narrow shingles are an important component of the built-up thatch look.

While there are no discernible rows in a built-up shingle thatch as opposed to a standard shingle roof, many of the same installation rules apply: all eaves have doubled, sometimes tripled courses; each shingle is fastened with two galvanized staples ¾ in. from the side edges and about 6 in. up from the butts; adjacent edges are gapped, although not as much as the ⅜ in. sometimes recommended; and joints between adjacent courses are usually offset by at least one inch (we bend this rule a little because of the number of courses in a built-up shingle thatch).

Almost all shingles are installed plumb, with butts square to the vertical fall line of the roof. This is more difficult than it sounds. On a steep pitch, material can be nailed mistakenly at a slight angle because the roofer is concentrating on a relatively small roof section, without the benefit of well-defined rows. Occasionally a nailer will have to scrutinize his work to regain an idea of what's "up" and "down."

The random pattern we apply in our thatching isn't as casual as it might appear. Many different nailers can be working on a project at once, and it's important that their work match seamlessly. From observing each other's shingling, nailers usually develop a sort of "mind meld," where individual differences in workmanship become impossible to detect.

In the last few years we've seen a growing interest in curved roofs, with architects and builders striving to design and construct eye-catching houses with radiused rakes and eaves and with virtually no joints at intersecting roof planes. As far as I can tell, there is no standard way to frame these curves (for one approach, see *FHB* #23, pp. 52-56). As for shingling them, if some crazed carpenter wants to jelly-roll sheets of plywood until his roofline looks like a roller coaster, with enough head-scratching we'll figure out a way to shingle it.

Eaves, gables, hips and valleys—The procedure for shingling eaves is basically the same whether they're square or radiused. For square eaves, we start by applying a double or triple course of shingles (depending on the desired appearance) over the elastic membrane and felt, making sure to offset the joints. If the eave is finished with a fascia board, we install our shingles with a mini-

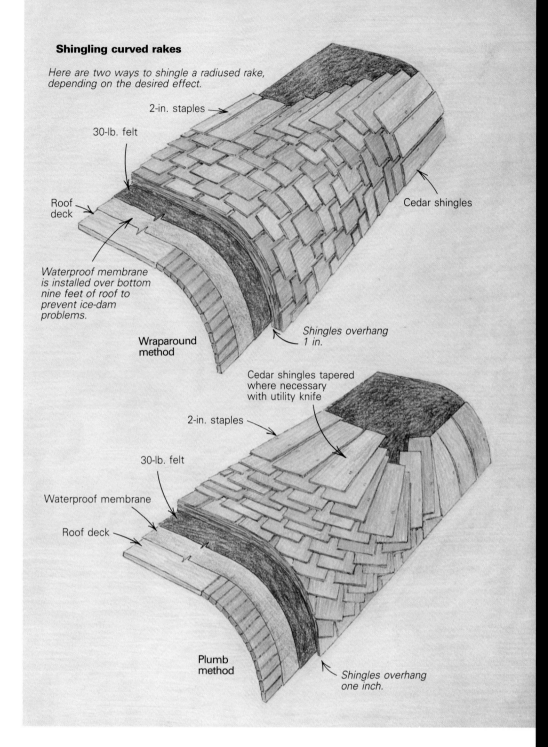

Shingling curved rakes

Here are two ways to shingle a radiused rake, depending on the desired effect.

2-in. staples

30-lb. felt

Roof deck

Waterproof membrane is installed over bottom nine feet of roof to prevent ice-dam problems.

Cedar shingles

Shingles overhang 1 in.

Wraparound method

Cedar shingles tapered where necessary with utility knife

2-in. staples

30-lb. felt

Waterproof membrane

Roof deck

Plumb method

Shingles overhang one inch.

mum 1½-in. overhang. If there is no fascia board—if a builder prefers, say, plaster soffits butted to the shingles—we allow more of an overhang to accommodate furring strips, lath and stucco (soffits are recommended because of the large quantity of staples penetrating the overhangs). Radiused eaves require the use of shingles pre-bent to the same radius as the eave. The number of curved shingles is dependent on the pitch of the roof (the greater the pitch, the fewer the number of curved shingles needed).

To enhance either radiused or square eaves, we sometimes install the starting course of shingles so that it undulates along the edge of a roof (top photo, p. 39). We do this by tapering the shingles with a utility knife and by "wave-coursing" the shingles in convex and concave arcs, making sure the apex of the arc overhangs the roof a minimum of 1½

in. The maximum overhang winds up being about five inches, which is possible because of the added strength and rigidity afforded by the small exposures. Usually builders don't want these arcs to be uniform, preferring an edge that appears to sag or droop in spots, so we lay the shingles free-style, with no layout beforehand. After this first decorative row is installed, we fill in the low spots with the usual random coursing.

Gable ends are also either square or radiused. On square rakes shingles are applied in the usual way, with a 1-in. overhang. For curved rakes, shingles are often wrapped around the bend of the roof, with the edges of the shingles parallel to the edge of the rake (top drawing above). Unless the radius of the framing is especially tight, the use of pre-bent shingles is unnecessary (remember we're dealing with narrow shingles that divide the radius down into small increments). The shingles are laid with a 1-in. overhang, the same as for a flat gable end. This is a fast, efficient method of shingling a curved rake, and we've done many of them this way. However, it's never appeared quite right to me because instead of the shingles being plumb along the radius, they're at an angle.

I prefer to fan the shingles down from the roof decking onto the radiused section so that the shingles remain somewhat plumb (bottom drawing, p. 37). This requires that the shingles making up the fan be tapered, that the bottoms of the shingles be cut at an angle along the edge of the roof to match the roof pitch and, finally, that a starter course be installed along the edge of the rake. It's a lot of extra effort, but I think it's worthwhile. To really confound matters, we sometimes lay the courses at the edge so that they undu-

late, too. In this case, we use the same procedure outlined above for the eave.

Hips and valleys, curved or not, are less complicated than curved gable ends. By weaving the shingles, we create the image of one roof section gently flowing into another. For hips, we accomplish this by tapering individual shingles to a particular angle (much like we do for arcs and curved rakes) and then combining them like pie slices until the proper turn has been made (bottom photo, p. 35). The angle of the taper depends on the roof pitch. For valleys, we simply open up the joints between adjacent shingles so that the shingles touch only at their butt ends (middle photo, facing page). Subsequent courses then cover the open joints. For both hips and valleys, a layer of Ice & Water Shield beneath the shingles serves as extra protection against moisture penetration. These same techniques allow us to roll a built-up shingle roof over virtually any rooftop structure, including all types of dormers and turrets.

Capping the ridge— For ridge caps, we use concrete barrel tiles, commercial cedar ridge cap or copper flashing (drawings left). Concrete tiles can be found in a variety of shapes and colors. We fasten them with 20d nails to allow for the thickness of the tile and the shingles, and we fill the gaps under and between the lapped tiles with mortar dyed to the appropriate color. While applying mortar, we cover adjacent shingles with polyethylene sheeting to protect them from staining.

Cedar ridge cap is made in lengths of 16 in. and 24 in. (we prefer 24 in.). This type of ridge tends to get lost on a massive thatch roof, so we recommend that adjacent lengths be lapped with about 5 in. to the weather (about one half the normal exposure), with the first course doubled on each end of the roof. This doubles the amount of ridge cap needed, but creates a bold, distinctive ridgeline. We fasten this cap with stainless-steel or galvanized 10d nails to assure adequate penetration into the sheathing.

Pre-bent copper (30 ga. or 28 ga.) also works well with a built-up cedar thatch, though it's more expensive then cedar. Our favorite design uses 10-ft. long sections bent in an "arrowhead" profile and fastened directly to the sheathing prior to shingling. Joints are lapped a minimum 4 in. and are caulked or soldered. The top few courses of shingles on either side are then cut progressively shorter so that their tops butt against the barb of the arrowhead. A bead of polyurethane sealant at the joint between the shingles and flashing prevents water from running under the shingles.

Flashing— Clients don't usually want to see lots of flashing on their roofs, so we're constantly weighing aesthetics against overall durability. Our real purpose as roofers is, after all, to make dwellings impervious to weather.

For plumbing and HVAC penetrations, the galvanized roof flashings usually supplied by subcontractors are acceptable. We install

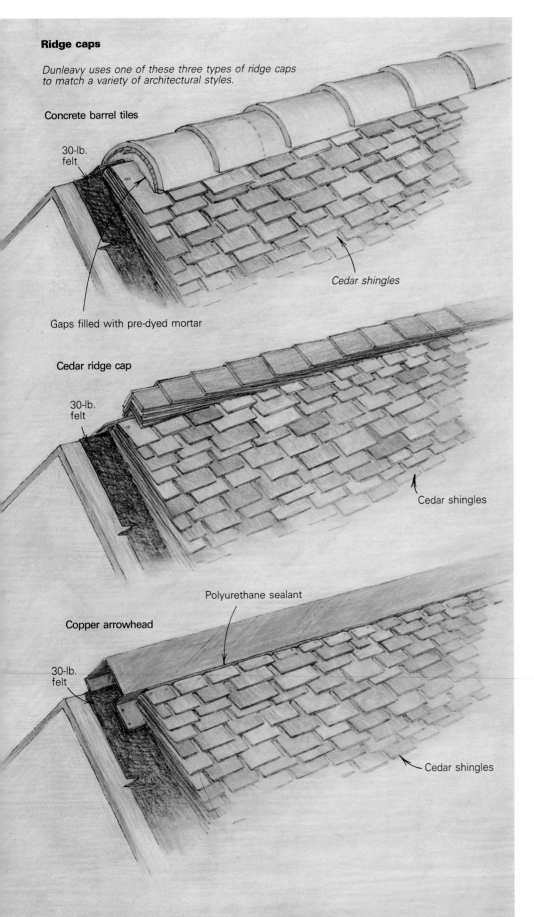

Ridge caps

Dunleavy uses one of these three types of ridge caps to match a variety of architectural styles.

Concrete barrel tiles

30-lb. felt

Gaps filled with pre-dyed mortar

Cedar shingles

Cedar ridge cap

30-lb. felt

Cedar shingles

Polyurethane sealant

Copper arrowhead

30-lb. felt

Cedar shingles

Instead of radiused rakes and eaves, some roofs terminate at square roof edges. As shown in the photo above, wave-coursing was used for the starter courses to create an authentic droopy appearance.

them much as we would on a standard-shingle roof. For sidewall flashing, we prefer a continuous piece of pre-bent copper or galvanized sheet metal over Ice & Water Shield and 30-lb. felt. We fasten each shingle with the staples placed as far away as possible from the flashing and maintain a 1-in. wide channel between the shingles and the wall to allow for water run-off. We avoid using step flashing because it's almost impossible to install properly on our built-up shingle roofs.

Because there are so many thicknesses of shingles in a built-up shingle thatch, we can hide the base flashing of a wall, skylight or chimney and still get a waterproof joint (bottom photo). To accomplish this, we run three or four full-length courses of shingles under the flashing and three or four progressively shorter courses over it. We're careful to place the staples toward the butts and at an angle to avoid penetrating the flashing. A high-quality caulk or flanged counterflashing is used to seal the joint where the shingles butt into the flashing. At the corners, we leave a small gap between the shingles. This method allows for drainage around any flashed component.

Roof-to-masonry joints can be handled the same way, although at times inappropriate counterflashing has been installed prior to our starting a project, forcing us to improvise. A classic example was a granite chimney and built-up cedar thatch, both intended to last a century or more, flashed with a thin piece of roll aluminum that might last 10 years.

Thatch roofs are closer to stonework than to roofing and hark back to an era in which a craftsman's work might well outlast him. As always, planning is the name of the game. ☐

Steve Dunleavy is a roofing contractor and owner of Hi & Dry Roofing in Stateline, Nevada. Photos by author except where noted.

As with hips, shingles are installed over valleys in continuous courses, creating a smooth transition between adjacent roof sections. Valley shingles are fanned out so that the shingles touch only at their butt ends. Shingles are affixed with the aid of pneumatic staplers.

The multiple layers of shingles in a built-up shingle thatch roof allow base flashings for skylights, walls and chimneys to lie hidden beneath two or three shingle courses, creating a more pleasing appearance. For the skylight pictured at right, a continuous bead of urethane caulk between the tops of the shingles and the base will keep water out. The side flashings are continuous because of the difficulty of step-flashing this type of roof, which is typically eight or nine courses thick. Water runoff is encouraged by holding shingles an inch away from the sides of the skylight, creating a gap between the shingles at the bottom corners.

Roofing with Asphalt Shingles

There's more to laying three-tab shingles than just nailing them on as fast as possible

by Todd A. Smith

I guess I was destined to work for my father's 63-year-old roofing company. I spent a lot of time on the roof learning the trade from several old craftsmen. I learned slate roofing, tile roofing, copper roofing and, of course, asphalt-shingle roofing. The most important lesson I learned was that there's more to installing shingles than just nailing them on fast. As a roofer, I am also charged with preventing leaks, making a house more attractive and re-

membering that everything a person has worked hard for is under my roof. In this article I'll describe the basics of installing a tight, durable three-tab asphalt-shingle roof.

Tools — You don't need many tools to install a shingle roof. A hammer, tin snips, utility knife, tape measure and chalkline will do it. I use a drywall hammer to nail shingles. Its light weight doesn't tire my arm, and the larger

head makes nails an easy target. To carry nails and tools, I wear a leather carpenter's apron, with suspenders to support the weight. Cloth aprons seem to wear out too easily. Besides, I like the extra pockets in my leather apron for different nails.

I prefer the snips when I have to cut shingles that butt into flashing or siding because I can make more accurate and intricate cuts with them. Otherwise, I use a hook blade in

With a column of shingles run up the center of the roof to the ridge, one roofer works to the left of center, the other to the right. This roof is shallow enough to walk on, but the roof brackets and scaffold provide a little extra room to stand and a place to rest bundles of shingles. Looped around the brackets, lengths of chain with hooks on the end make a handy place to hang a nailer or a bucket full of nails.

my utility knife because it's less likely than a straight blade to cut whatever is underneath the shingle. When cutting a shingle on the roof, I use the back of another shingle as a straightedge, which saves me from having to carry a square.

Our company's air-powered nailers speed up production, but they have their drawbacks. Nailers are heavy, and dragging 100 ft. of hose is cumbersome. Our nailers hold only enough nails to install one bundle of shingles, which means we have to keep coils of nails on the roof, and they get in the way. Also, in areas that require lots of cutting and fitting of shingles, power nailers are clumsy and impractical.

To make the nailers easier to handle on the roof, I made some portable utility hooks that attach to our scaffold brackets. This gives me a place to hang my nailer when I'm not using it (photo facing page), as well as a place to hang buckets of nails and coils of air hose. The hooks are made with loops of chain, spring latches and some utility hooks, all of which I got at a hardware store.

Ladders and roof brackets—For us, safety begins as soon as the trucks show up at the job. Roofing can be hard on your back, and you have to be careful just pulling the ladders off the truck and setting them up, not to mention hauling heavy rolls of felt and bundles of shingles up on the roof.

I've become accustomed to setting up extension ladders alone because I have to examine many roofs by myself. I slide the ladder off the back of the truck until its feet touch the ground, then tip it up on one edge. Squatting under the ladder with a little over half its length in front of me, I stand up, placing the ladder on my shoulder, keeping my back straight and lifting with my legs. With the majority of the weight in front of me, I don't have any downward pressure behind me to strain my back.

To stand the ladder up, I set the foot of the ladder against a solid object—usually a foundation, step or tree trunk—and push on the top, walking my hands down the rungs. Once the ladder is straight up, I raise it to the appropriate height, watching out for power lines, phone lines and tree branches.

There are two accessories I use with my ladder when the needs arise. One is a ladder standoff, which is a large U-shaped affair that bolts to the ladder and prevents it from leaning directly against the eaves of the house. The other is a ladder scaffold, which is a platform supported by two brackets that hang over the rungs on a pair of extension ladders.

I use the ladder scaffold to work along the eaves of a house. My steel scaffold platform is 20 ft. long, and to be sure it overhangs the brackets at least 12 in., I stand up two ladders about 17 ft. apart. Once the ladders are up, I set the brackets at the proper working height, which for me is about 3 ft. below the eaves. The brackets are adjustable and can be oriented in a horizontal position no matter what the ladder's angle. Because of its size and weight,

getting the 20-ft. scaffold up the ladder and onto the brackets is a two-man job, and once in place, I tie the scaffold to the ladders.

About 90% of the time I'm working on a roof that's too steep to walk on (anything greater than a 6-in-12 pitch), so I set up roof brackets (photo facing page). These are triangular steel brackets, some of which can be adjusted level regardless of the roof pitch. Others are fixed and create a working surface that allows me to walk around on the roof easily (for more on scaffolding see *FHB #36* pp. 34-38).

The basic materials—Asphalt shingles and fiberglass shingles are the most common roofing materials used today. Both are less expensive than slate, tile or metal roofing and are more fire resistant and maintenance-free than wood. The main difference between asphalt and fiberglass shingles is in the mat, or base sheet, that manufacturers begin with when they make shingles. Asphalt shingles have an organic base, which, like felt underlayment, is

Estimating roofing

Roofing shingles are sold by the square: a square of shingles is enough to cover 100 sq. ft. There are 27 shingles in a bundle of standard three-tab shingles, and three bundles to a square. I calculate the square footage of the roof and divide that figure by 100 to determine how many squares of shingles I need. For every 5 ft. of hip or ridge, I need four shingles. I generally figure 30 ft. to 40 ft. of hips and ridges for every square of shingles. When a roof has many hips, valleys and irregular shapes, I figure 10% to 20% extra for waste.

From the blueprint or specs, I also determine if I will need drip edge, flashing or any other additional material. The drip edge and flashing are linear dimensions. I figure 2 lb. of nails for each square of shingles. I use 1¼-in. galvanized nails for new construction. I don't use staples to install shingles because I feel the head of a nail holds the shingle on better. Manufacturers of roofing felt assume a 2-in. overlap, so a roll of 15-lb. felt that will cover 400 sq. ft. of roof has 432 sq. ft. in the roll.

The cost of installing the roof is the sum of the materials, labor, overhead and profit. Most roofers determine the labor cost based on a set price for each square of roofing. The problem with this is that not every square of roof shingles takes the same amount of time to install. A square of roofing material on a shallow-roofed ranch house will take less time to install than a square of roofing three stories high on a steeply pitched Victorian. Instead, I break down the roof into sections and determine the time each one will take a specific roofer or group of roofers to complete. I am always conscious of details that require extra time, like valleys and flashings. —*T. S.*

composed of cellulose fibers saturated with asphalt. Fiberglass shingles have a base of glass fibers and don't need to be saturated with asphalt.

Beyond that, asphalt and fiberglass shingles are made pretty much the same way. The mat is coated with asphalt on both sides, and then ceramic-coated mineral granules are applied. The granules help shield the asphalt from the sun, provide some fire resistance and add color. The seal strip, applied along the width of the shingle and just above the cutouts, is activated by the sun and seals each shingle to the one below it.

On a standard three-tab shingle, the top edge of each shingle is marked with a notch every 6 in. The notches are used to register the shingles in the next course, since alternate courses are offset from each other by 6 in. Some manufacturers even notch the shingles every 3 in. to allow for a pattern that repeats every third row as opposed to every other row.

Fiberglass shingles are a little more difficult to install than asphalt. In hot weather, they become softer much faster, which makes them harder to handle and cut, and easier to damage.

Fiberglass shingles are lighter in weight, though. A square (100 sq. ft.) of standard three-tab fiberglass shingles weighs 225 lb. A square of asphalt shingles weighs 240 lb.

The two most common types of roofing felt are 15-lb. felt and 30-lb. felt. The designation is based on the weight of a 100-sq. ft. area of felt. Although some people don't bother to install felt under the shingles, we always do. For one thing, felt protects the roof decking from the weather. Even though roof decks are made from exterior-grade plywood, they won't withstand prolonged exposure to the elements. Second, roofing felt helps prevent ice and snow from backing up under the shingles and leaking water into the house.

Installing felt and flashing—Felt is applied in courses, parallel to the eaves. Generally courses overlap each other by 4 in. The rolls are marked with white lines along the edges to help you maintain a consistent overlap. There's also a pair of lines in the center of the felt, in case you want to overlap the courses 18 in. (half the sheet). You might want to do this on a shallow-pitched roof, say 4-in-12, in an area prone to ice damming.

We try to run the length of the roof with a piece of felt, trimming it flush with the gable ends. But if we have to splice in the middle of a course, we overlap the ends 4 in. We run felt 6 in. over all hips and ridges (from both sides). Valleys are lined with a full width (36-in.) piece of felt first, and then the courses are run into it, overlapping the sides of the valley felt by 6 in. Where a roof butts into a sidewall, we run the felt 4 in. up the wall.

When we come to a vent pipe, we cut a 3-ft. piece of felt, make a hole in it the size of the pipe, slip it over and seal around the pipe with roofing cement. Then, we overlap the felt on both sides of this piece. To avoid having to chalk horizontal lines later as a guide for the shingles, we install the felt as straight as possible so that we can measure off of it to keep the shingles straight.

We nail along the seams and edges of the felt. In the center of each course, we nail every 2 ft. or so. Although you can nail by hand, I usually use a power nailer filled with 1¼-in. roofing nails, which are the shortest pneumatic nails I can get. When I need to install a small area of felt quickly and don't want to bother with a compressor and hoses, I use an Arrow Hammer-Tacker (Arrow Fastener Company, Inc., 271 Mayhill St., Saddle Brook, N. J. 07662). It's a staple gun that is used like a hammer, but tends to gum up with felt after a lot of use. When that happens, we soak the heads in kerosene to break down the tar. Then we scrape them clean and spray them with lubricant.

Once the felt is on, we nail strips of lath along the edges of the roof and along the seams in the felt to prevent the felt from blowing off. Until the roof is done, the felt is the only material keeping the house dry.

The edge of the roof sheathing should always be protected from the elements. I use a metal drip edge, installed with roofing nails, along the gables and eaves. At inside corners, I cut the vertical face of one piece 1 in. long and bend it around the corner, and then cut the other piece so that it butts into the corner. I cut the tops long on both pieces and overlap them in the valley. At outside corners on a hip roof, I cut a "V" out of the top section and simply bend the drip edge around the corner. Gable ends are cut flush with the rake board at the bottom. At the peak, I cut a "V" out of the vertical flange and bend the top section over the ridge. Any splices in the drip edge are overlapped 2 in.

Houses without overhanging eaves are particularly susceptible to damage from ice damming. On such houses we also install a 36-in. wide strip called an ice shield, or eave flashing. Although you can use roll roofing, we use any of several membrane products specifically designed for this like Ice & Water Shield (W. R. Grace & Co., 62 Whittemore Ave., Cambridge, Mass. 02140) or Weather Watch (GAF Corp., 1361 Alps Rd., Wayne, N. J. 07470-3689). We roll them out along the eaves, tacky side down, and nail them only across the top.

Loading the roof—Shingles should be stored in the shade, or covered with a light-colored tarp. Otherwise, the sun will heat them up and seal them to each other. For this reason we seldom unload shingles directly onto the roof, even though some suppliers have lift trucks that make it possible. We either carry the bundles up by hand, or we use a gas-powered hoist (ours was made by Louisville Ladder Division, Emerson Electric Co., 1163 Algonquin Pkwy., Louisville, Ky. 40208).

Some roofers carry the shingles up to the ridge and lay the bundles over it. This isn't a good idea, especially with fiberglass shingles. The shingles can heat up, take a set

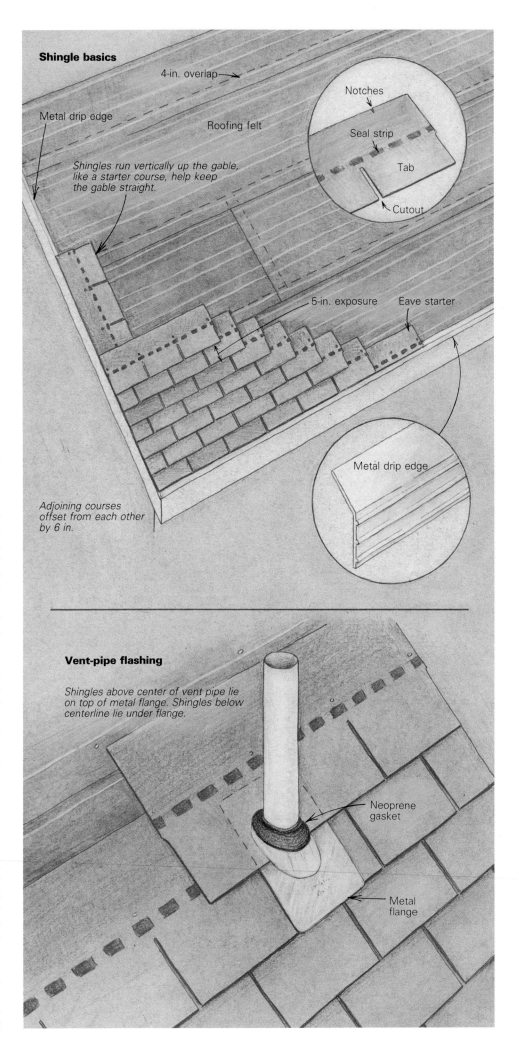

Shingle basics

4-in. overlap

Notches

Metal drip edge

Roofing felt

Seal strip

Tab

Shingles run vertically up the gable, like a starter course, help keep the gable straight.

Cutout

5-in. exposure   Eave starter

Metal drip edge

Adjoining courses offset from each other by 6 in.

Vent-pipe flashing

Shingles above center of vent pipe lie on top of metal flange. Shingles below centerline lie under flange.

Neoprene gasket

Metal flange

Drawings: Chris Clapp

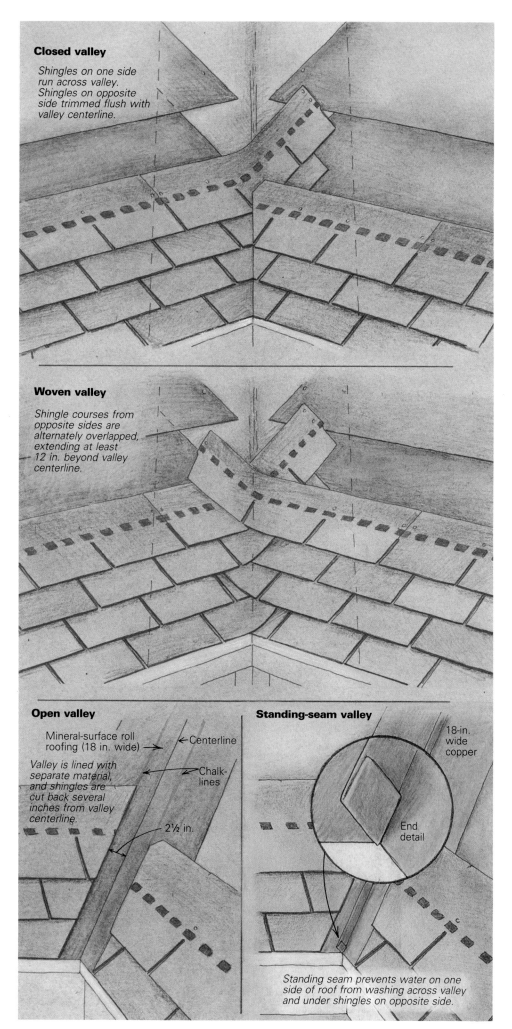

Closed valley

Shingles on one side run across valley. Shingles on opposite side trimmed flush with valley centerline.

Woven valley

Shingle courses from opposite sides are alternately overlapped, extending at least 12 in. beyond valley centerline.

Open valley

Mineral-surface roll roofing (18 in. wide) →   ← Centerline

Valley is lined with separate material, and shingles are cut back several inches from valley centerline.

← Chalk-lines

2½ in.

Standing-seam valley

18-in. wide copper

End detail

Standing seam prevents water on one side of roof from washing across valley and under shingles on opposite side.

from being bent and then not lie properly. We spread the bundles out around the roof, on approximately 6-ft. centers. Any farther apart and you'll work yourself to death. Even on shallow roofs, the bundles can slip on the felt and slide off, so we drive a pair of 10d nails into the sheathing and set the bundles above them. The nails can make a dimple in the edge, especially in hot weather, so we always lay the bundles with top edge of the shingles against the nails.

Starting the shingles—Before the first course of full shingles goes on, starter shingles are installed along the eaves. Their purpose is to protect the roof under the cutouts and joints in the first full course and to provide a seal strip for the tabs. Although some people use whole shingles installed upside down as starters, I use full length shingles with the tabs cut off (top drawing, facing page). This puts the seal strip in the right place to hold the tabs.

Along the gable end, I like to let the shingles overhang the drip edge by 1 in. This keeps water from blowing under them. To keep this line of shingles ends straight, I run a course of shingles vertically up the gable, like a starter course (top drawing, facing page). I lay them end to end from eaves to ridge, with the cutouts toward the roof and the top edge overhanging the drip edge about one inch.

Once the starter course is in place, the field shingles can be started. I begin in the center of the roof and work outward, if there is a large open area, or if I'm on a hip roof. I begin on the gable end when there is no large open area to nail shingles, or when there are dormers in the way and I won't be able to get a proper chalkline on the roof.

I install the first five or six courses without chalking any lines, following the eaves with the first course, and then following the notches and cutouts of the previous course. If I need to install roof brackets, I usually do so after the sixth course. This puts them at a comfortable height from which to get on and off the ladder. Also, I can nail two or three more rows of shingles beyond the roof brackets while still standing on the ladder, which is easier than nailing these rows from the scaffold.

Once you're up on the roof, you can work horizontally across the roof, installing several courses as you go. Or you can work vertically up the roof, installing the first two or three shingles in each course all the way to the ridge. I prefer a combination of the two. On our crews one roofer sets the pattern by nailing vertically up the roof (in the center when possible). The other roofers nail horizontally, one to the left and one to the right (photo, p. 40). This increases productivity by allowing three roofers to install shingles on the same roof without getting in each other's way.

We chalk lines on the roof to help us keep the shingle courses aligned and straight. I start with a pair of vertical lines, one in the center of the roof and another 6 in. away (on either side). All subsequent courses are be-gun from these lines.

If I'm starting from the gable end rather than in the center of the roof, I don't chalk any vertical lines as long as the gable is straight. Before I go up on the roof, I make a pile of full shingles and a pile of shingles that have 6 in. cut off of them. Then when I'm on the roof, all I have to do is alternately pull shingles from each pile and work my way up the gable. Successive courses are automati-cally offset by 6 in.

When you reach the opposite gable with a course, the shingles are simply trimmed flush with the starter course running up the gable. Rather than trim the last shingle in each course individually, some roofers let these shingles run long and trim them later. If the shingles are not cut as they're installed, I insist that they be cut at least every few hours. If left any longer, they will droop against the side of the house, making it diffi-cult to get a straight cut.

Unless I installed the felt and know that it's straight, I chalk horizontal lines every six to eight courses. (Roofing felt is often installed by the frame carpenters before we get to the job.) I measure 5 in. for each course, which is the amount of shingle exposure (this varies with the manufacturer, so be sure to check this by measuring the depth of the cutout on your shingles). I periodically measure from my chalklines to the ridge to make sure I'm run-ning parallel to it. If not, I adjust the chalk-lines in small increments.

These horizontal lines help align the top edges of the shingles. Yet because individual shingles may vary in height as much as ¼ in., aligning the top edge doesn't guarantee a straight course. But chalking lines for the top edge means that we can chalk all the lines at once, which saves us time.

If the courses are running off and we need to straighten them out, we'll chalk a line along the tops of the cutouts of the last course installed. This aligns the bottom edges of the shingle and results in a perfectly straight course. If the shingles are running way off, we'll straighten them out gradually over several courses. We always use blue chalk, which washes away in the rain. Red chalk can permanently stain the shingles.

I use four nails to a shingle—one at each end and one over each cutout. On steep roofs I may install a fifth nail at the top center of the shingle. Contrary to conventional wis-dom, I always nail right above the seal strip, not on it or below it. Nailing on the seal strip prevents proper adhesion. Below the seal strip, the nails are too close to the edge and can rust or corrode, causing the shingles to slip and the roof to leak. I have never had a shingle blow off or slip because I nailed above the seal strip.

Field shingles that end up over a hip are just run long, nailed and then cut down the center of the hip with the hook blade of a utility knife. At the ridge, we trim one side of the roof flush and fold the shingles from the

Working his way up an open valley lined with mineral-surface roofing, Smith uses tin snips to trim shingles flush with the chalkline. Blue chalk is preferred because it washes away in the rain; red chalk will permanently stain the shingles.

other side over the top. This is just a tempo-rary measure to seal the ridge until the caps are installed.

Valleys—There are three different types of valleys used on shingle roofs: the closed val-ley, woven valley and open valley (drawings previous page). A closed valley is created by running shingles from one side of the roof across the valley (at least 12 in. beyond the centerline), and then overlapping them with shingles from the other side, trimmed flush with the centerline of the valley. It is the cheapest, easiest and least durable option.

The woven valley is created by alternating and overlapping each row of shingles from the left and right sides of the valley. The shin-gles should overlap the valley by at least 12 in. and nails should be kept at least 8 in. away from the valley centerline.

An open valley is created by lining the val-ley with a separate material (roll roofing or copper, for instance), and then cutting the shingles back so the lining materal remains exposed. We use open valleys on 90% of our roof installations because they are the most durable. An open valley allows water to flow easily off the shingles. Depending on the budget, the type of shingle and the style of the house, we'll line open valleys with miner-al-surface roll roofing, copper or lead-coated copper. One material we never use for val-leys is aluminum because it expands and contracts more than other materials, which causes it to wear out much faster.

For mineral-surface valleys, I use strips of material 18 in. wide and 8 ft. to 10 ft. long. A longer piece is difficult to work with and may rip during installation. It comes in 36-in. wide rolls, which I cut down the middle. I cut the valley material on the ground, using a utility knife, and loosely roll it up to make it easier to carry on the roof. Some manufacturers produce roll roofing in colors to match their shingles. But both black and white roll roof-ing are pretty common and look good with most shingles.

To install the valley, I start from above and unroll the material. I center it, nail one of the upper corners, then work my way down, nail-ing the same side every 12 in. to 18 in. It is imperative for the valley material to be tight against the sheathing. If it is not, it can break under the weight of ice and snow. To avoid misalignment, only one person should nail the valley, always working from top to bot-tom. First one side, and then the other. When I need a second piece to complete the valley, I overlap the first piece 4 in., but don't usual-ly seal between them.

Metal valleys are installed pretty much the same way. Again, I use 8-ft. to 10-ft. lengths. I like to crease the center of the valley metal on a sheet-metal break. The job looks neater and cleaner this way.

I use a standing-seam valley when a steep roof drains onto a flatter shingle roof. I bend the valley metal on a sheet-metal break, cre-ating a ridge in the middle of the valley (bot-tom right drawing, previous page) that pre-vents water from the steep side of the roof from flowing across the valley and running up under the shingles of the shallower roof. To finish off a standing-seam valley at the bottom, I cut the sides flush with the eaves, but let the standing seam run about 1 in. long. Then I simply fold the seam back on itself and crimp it tightly. At the top, I start about 6 in. from the ridge and bend the standing seam over with a rubber mallet so that it lies flat across the ridge.

It is important to seal valleys at the top to keep water from getting under them. With a mineral-surface valley, I use a 4-in. wide piece of roofing fabric and apply one coat of roofing cement under it and another coat on top of it. Roofing fabric is a cheesecloth-like material or fiberglass mesh, saturated with

Using a hook blade in a utility knife, Smith cuts ridge caps on the ground (photo left), tapering the tops so that, once installed, the lap portion will be neatly hidden beneath the exposed portion of the succeeding shingle. The caps are centered on the ridge and held in place with one nail on each side (photo above). The last cap on the ridge will have the lap portion trimmed off, and because the nails will be exposed, they'll be sealed with caulk.

asphalt. It comes in 4-in., 6-in. and 12-in. wide rolls. The 4-in. rolls cost about $7 for 100 ft. of fabric and are available at most building-supply stores.

When I don't have anything to solder to, I use the same technique to seal the tops of copper valleys. However, when two copper valleys meet at a ridge or when a copper valley meets copper flashing, I prefer to solder them. To solder two valleys at a ridge, I end one valley at the ridge and bend the other valley over the ridge about 1½ in. (again using a rubber mallet).

After installing the valley material, I chalk lines along both sides of the valley, 2 in. to 2½ in. from the center. These are the marks I follow to install the shingles. The exposed sections of my valleys are between 4 in. and 5 in. wide total. Anything smaller is impractical and anything larger doesn't look good.

I prefer to cut the shingles even with the chalklines while they are being installed, using a pair of tin snips (photo facing page). I nail the full shingle in place, use another shingle as a straightedge, scribe a line with the point of a nail, then bend the shingle up and cut it with my tin snips. I can install and trim the valley shingles on both sides as I work up the valley to the ridge.

Some roofers let all the shingles run long. As each side is completed, they chalk lines down the valley and use a utility knife to cut the shingles (being very careful not to cut the valley material). With either method, it's important to keep the nails 4 in. to 6 in. away from the edge of the shingles in the valley.

Flashing—I use step flashing along the sides of a wall or chimney (wood or masonry). Step flashing consists of small squarish pieces of metal, bent in an L-shape. The individual pieces are installed with each course of shingles so that the shingles in the succeeding course hide the exposed metal (for more on installing step flashing see *FHB #35* p. 50). You can buy precut pieces of aluminum step flashing, but I prefer copper, so I have to make my own from 18-in. wide rolls. I make my steps at least 9 in. wide by 8 in. long, which is equivalent to 5 in. of exposed shingle and 3 in. of headlap. I nail the pieces of step flashing on sidewalls before the siding is installed. Then the siding acts as the counterflashing. In the rear of a chimney, I install a copper cricket, which is a saddle that diverts water around the chimney (see the article on pp. 98-100).

To flash around soil and vent pipes on most houses, we use a manufactured metal flange with a neoprene gasket (bottom drawing, p. 42). It's important to remember that all soil and vent-pipe flashings lie on top of the shingles from the center of the pipe forward and lie under the shingles from the center of the pipe back. We shingle up to the center of the vent pipe, either by notching the shingles around it, or by actually cutting a hole in one of the shingles and slipping it over the pipe. Next we slip the metal flange over the pipe and nail the top corners. Then we continue applying shingles, notching them around the pipe and being careful to keep nails away from the pipe (for more information on flashing techniques, see *FHB #9* pp. 46-50).

Capping hips and ridges—Hips and ridges are capped with shingles that are only one tab wide. The caps are made by cutting three-tab shingles into three pieces (photos above). Once again, this is done on the ground. The shingles are cut with a utility knife, starting at the top of the cutouts and angling slightly inward, toward the top of the shingle. This assures that the lap portion of the shingle will be neatly hidden beneath the exposed portion of the succeeding shingle (photo above right).

Of all the areas where I have installed shingles, the hips are the most difficult to keep straight without a chalkline. I snap a line parallel to the hip, about 6 in. away (it doesn't matter on which side). This acts as a guide for the outer edge of the hip caps. I start at the bottom, cut the first cap even with the eaves shingles, then I work my way to the ridge.

I nail hip and ridge caps on or just above the seal tab. Nailing higher will cause the bottom of the shingle to pop up. Hip and ridge caps are nailed with about a 5-in. exposure. The bottom of each cap is aligned with the top of the cutout on the previous cap.

Most roofers install ridge caps from one end of the roof to the other, orienting them so that the prevailing winds blow over the caps, not under them. I prefer to work from both ends toward the center. I can't really say why, except that it's the way I was taught. On steep roofs, where the centerline of the ridge is more distinct, I don't usually chalk a line for the ridge caps. On shallow roofs, I do. Once both sides of the ridge reach the center, the last ridge caps have to be trimmed so that they butt together, otherwise the ridge will have a lump in it at this point. One last cap piece will cover the butt joint. It should be only about 5 in. long and installed with two exposed nails, both of which are caulked with clear silicone caulk.

Where the ridge of a dormer meets a roof, I work from the front of the dormer back toward the roof to install the caps. The last cap spans across the seam between the valleys (which is sealed with solder or roofing fabric). And the field shingles on the roof lie over the last dormer ridge cap. □

Todd A. Smith is a roofing contractor in Verona, New Jersey. For more on asphalt-roofing techniques, contact the Asphalt Roofing Manufacturers Association (6288 Montrose Rd., Rockville, Md. 20852) for a copy of their Residential Asphalt Roofing Manual, *$10.*

Roofing with Slate

Shingling with stone will give you a roof that lasts for generations

by David Heim

When my wife Katie and I bought a farmhouse in northeastern Pennsylvania a few years ago, we weren't surprised that it had a slate roof. Built sometime before 1860, the house is close to several slate quarries. Here and in other parts of the Northeast (photos below), slate roofs are common on old houses.

After more than 120 years, the original slate on our roof was suffering from age and neglect. But it was one of the most attractive features of the house, so our plan from the outset was to repair or replace it. The roof didn't leak when we bought the house, although it had in the past. Previous owners had smeared tar in between the slates in several places, and some of the slates had been replaced with pieces of tin. Many slates were soft to the touch and crumbled readily.

Over time, even slate yields to the intrusion of water; and once this happens, freeze-thaw cycling causes it to delaminate along its cleavage planes (sidebar, facing page). Cracks in the surface (called crazing), flaking (spalling) and chalky deposits around the edges and unexposed face of the slate are signs that slate has reached the end of its useful lifespan.

We could have patched the roof to make it last a few more years, but we decided instead to add a new roof to the list of improvements we had planned for the house. Our first inquiries into re-roofing with slate didn't bring very positive responses. Slate is prohibitively expensive, we were told; it's too hard to work

with, and there aren't any good slate roofers around any more.

After some persistent investigating, though, we realized that what we'd been told wasn't entirely true. In my area, slate isn't much more expensive than other good roofing materials—particularly not in the long run. And we found a contractor, Jim Hilgert, who knows slate work backwards and forwards. Having seen how Hilgert's crew handled our roof, I'm convinced that roofing with slate isn't much harder than roofing with other shingles. It's in cutting, hole punching and handling that slate work differs. Though the job shown here is a re-roof, you would use the same tools and techniques to put on a new slate roof.

Selecting the material—Slate comes in a wide variety of colors, depending on where it is quarried. For our re-roofing job, Hilgert used #1 clear Pennsylvania blue-grey slate, salvaged from a barn in New Jersey that was about to be torn down. The pieces were 24 in. by 12 in. (with an exposure of 10½ in.), the same size as our originals. The slate was in good condition and only about 20 years old—

not yet middle-aged, as slate goes. Pennsylvania slate is reputed to last for at least 75 years. Slate quarried in Vermont and Virginia can last 150 years or more on a roof.

Including removal and transport, we paid $100 a square (enough to cover 100 sq. ft.) for our slate, or about half the local quarry price at the time. Vermont slate can sell for up to $300 a square, but even at that price it's not a bad deal when you consider the longevity of the material. Fiberglass shingles, which cost about $60 a square in our area (including roof sheathing and felt underlayment), would have to be replaced three or more times within the lifespan of a single slate roof.

Recycled slate isn't always the bargain it appears to be. Usable old material can often be impossible to find, even in an area where slate is commonly used. For example, four months after Hilgert bought our slate, he was unable to find more second-hand slate for another job. When you do come across salvageable material, it takes time and experience to cull out the bad slate.

To be sure that none of our slates had hairline cracks or delaminations, Hilgert "rang"

the slates as he pulled them off the barn roof, much as you'd ring a china cup to check its soundness. If you hear a faint echo when you tap the slate—something like the sound your knuckles make when they rap a solid plank of wood—the slate is all right. A dull thud with no resonance indicates unsound stone that's best rejected. Hilgert also rejected "ribbon" slates—pieces that have a pale streak running through them. This impurity in the stone is a weak spot that won't weather well and will crack prematurely. Slates without ribbons are called "clear."

Once he'd found enough material, Hilgert hosed the slates down to remove the accumulated stone dust. Dipping salvaged slate in a solution of oxalic acid and water (wear rubber gloves) will remove weathering marks and restore the slate surface to good-as-new condition, but it takes a lot of time. Hilgert trucked the cleaned slates to our house and stacked them on edge, like large, thin dominoes.

Preparing the framing—Like many other houses in our area, our roof has almost no sheathing. The slates were fastened to roof

From quarry to roof

In terms of composition, slate is little different from the clay deposits you might find in a river bed. It's the geological forces of pressure, temperature and time that transform clay into shale and slate. Both are sedimentary rocks, but shale is softer and less dense because it hasn't been cooked or compressed as much as slate. When slate forms, tremendous temperatures and pressures cause the mineral grains to align so that they're parallel to each other. This granular alignment creates the cleavage planes that enable quarry workers to split out thin, flat sheets of stone.

Splitting slate along its cleavage plane reveals the surface texture, or grain, of the slate. On premium-quality slates, the grain should run lengthwise, as it does on a cedar shingle. Grain can vary from smooth to coarse, and a rough surface doesn't mean that the slate is poor-quality material. Smooth slates are easier to work with, however.

Slate color depends on chemical and mineral makeup, and can vary from the grey stone quarried in eastern Pennsylvania to the red and green tones found along the Vermont-New York border. Other standard colors established by the Department of Commerce are black, blue-black, blue-grey, purple, mottled green and purple. *Ribbon* slates are streaked because of impurities in the original clay deposit. In some cases, this ribbon can weather prematurely, so slates classified as *clear* are a safer bet for a long-lasting roof. Color is further

qualified as either *unfading* or *weathering*. Some slates change color over time, but those designated as unfading will not.

Standard roofing slate is 3/16 in. thick and can be ordered in a number of sizes, from 10 in. by 6 in. to 24 in. by 14 in. These slates are fairly uniform and usually have their holes (two per slate) machine-punched at the quarry. To install what is known as a *textured* slate roof, you'd use slates that vary in thickness from 3/16 in. to 3/8 in. The *graduated* slate roof is another variation involving slates of different sizes and thicknesses. Usually the larger, thicker (sometimes up to 2 in.) slates are located near the eaves, with thinner slates and less exposure used near the ridge. These roofs allow considerable aesthetic expression on the part of the slater, and no two are the same, as the photos at left, taken in New England, show. Most of the slate work done today, however, is with standard slates.

Ordering slates—Like other roofing materials, slates are sold by the square. A square of slates should cover 100 sq. ft., with the standard 3-in. lap. Slate size determines the number of slates in a square, and the exposure to the weather. Exposure is easily figured with a simple formula: Subtract 3 in. from the length of the slate, and divide by 2. The 24-in. by 12-in. slates used for Heim's roof come 115 to the square; 12-in. by 8-in. slates come 400 to the square.

Slate prices can vary a great

deal, depending on size, thickness and color. Quarry prices start at $300 to $400 per square. Unless you're near a supplier (see the list of operating quarries below), freight charges may end up determining what your best delivered price is. Most quarries don't have a full range of sizes in stock. Special orders can be cut, but you'll have to wait for them. And remember that the smaller size slate you use, the longer it will take to nail up. Larger slates—18 in. or longer in standard or random widths—can really go up quickly. What this boils down to is that a little phone work can go a long way toward saving time and money.

If you're new to slate roofing, there's a good book available from Vermont Structural Slate Co., Inc. (Box 98, Fair Haven, Vt. 05743; $7.95 postpaid). Entitled *Slate Roofs* and originally published in 1926, the book provides a detailed, state-of-the-art look at slate work in its heyday.

Below are names, addresses and telephone numbers of four major slate quarries that operate on a year-round basis.

Buckingham Virginia Slate Corp., Box 11002, 4110 Fitzhugh Ave., Richmond, Va. 23230; (804) 355-4351.

Rising and Nelson Slate Co., West Pawlet, Vt. 05775; (802) 645-0150.

Structural Slate Co., 222 E. Main St., Pen Argyl, Pa. 18072; (215) 863-4141.

Vermont Structural Slate Co., Inc.; Box 98, Fair Haven, Vt. 05743; (802) 265-4933. —*Tim Snyder*

Traditional tools. The slater's stake is T-shaped, and its sharp end can be driven into a rafter or other wood work surface. Its horizontal edge supports the slate while it's punched, cut and smoothed. The hammer, which is made from a single piece of drop-forged steel, is designed to drive and pull roofing nails, to punch holes and smooth the rough edges of cut slate.

laths (purlins) of 4/4 by 2-in. hemlock that had been nailed across roughsawn 4x6 rafters.

The roof lath on our house is 10½ in. o. c., the spacing required to give our 24-in. by 12-in. slates a 3-in. lap (drawing, facing page). (Lap refers to the required triple overlay of slates on three consecutive courses.) Roofs pitched shallower than 6 in 12 should have a 4-in. lap. Very steep roofs, like mansards, can get away with a 2-in. lap.

Hilgert framed the roof of the new bathroom we added in the same manner as the house. Instead of 4x6 rafters, he used standard 2x8s; and for roof lath he used 4/4 by 2-in. white pine. Only along the ridge and the eaves do you have to sheathe the rafters. On our roof, Hilgert used wide 4/4 boards.

As a rule, you can get by with conventional framing if you're installing a standard slate roof like ours. But increasing rafter size by one nominal dimension (from 2x8 to 2x10, for example) would reduce the deflection of these members over the years, particularly with a snow load. We used standard ³⁄₁₆-in. slate, which weighs between 750 lb. and 850 lb. per square, depending on where it was quarried. If you're planning what's known as a textured or a graduated slate roof, you will need to beef up your framing considerably. These two roofing styles call for slate that's ⅜ in. to 2½ in. thick, which translates into loads of 1,500 to 6,000 lb. per square.

Though some slate roofers prefer to use conventional sheathing beneath a slate roof, we decided to stay with the original 4/4 lath, since most of it was in good shape. Hilgert also believes that an airspace directly underneath the slate allows it to dry out more thoroughly after a storm.

If you decide to install a slate roof over sheathing, the sheathing should be covered with overlapping layers of 30-lb. asphalt felt before the slate goes on. The felt protects the roof from weather while the slate is being laid, and also forms a cushion for the slates.

The rafters in the main part of our house were in excellent condition, and I knew that they could carry the weight of the new slate with no problem. After removing the old slate,

however, we found that some of the old hemlock roof lath would have to be replaced, along with a few of the wide boards at eaves and ridge. Some of this old hemlock had become so hard and brittle over the years that you couldn't drive a nail into it without causing entire runs of lath to vibrate. This, in turn, caused already installed slates to shake and pull free from their nails. So this old wood was replaced with new white pine.

Cutting and hole punching—The traditional tools for these tasks are a slater's hammer and stake (photo left). You probably won't find them at your local hardware store; I got mine at a flea market. New tools are available from John Stortz and Son, Inc., (210 Vine St., Philadelphia, Pa. 19106).

The hammer is made from drop-forged steel and has a leather handgrip. Between handgrip and head, the handle is flat, with one edge beveled sharp, so that the tool can be used to smooth rough edges of trimmed slate. Where the handle joins the head, there's a stubby pair of claws for pulling nails. The striking face of the head is small—about the size of a nickel—to minimize the risk of damaging the slate when nails are driven home. The other end of the head tapers to a fairly sharp point. This sharpened end is used to punch nail holes in the slate, and to perforate slate along a scribed cutting line. Once perforated, the slate can be broken, and the resulting jagged edge can be smoothed with the beveled edge of the handle.

The technique for punching and perforating takes time to master. It's a short, quick, well-aimed stroke that stops just after the hammer's metal point strikes stone. Smoothing a cut edge with the beveled handle is easier: just chisel the slate smooth. It's not a bad idea to practice your technique on a few broken pieces of slate before working on slate to be nailed up.

An alternate hole-punching method is to drive a nail through the back of the slate. Always work on the face that won't be exposed to the weather. This way, the slightly broken or beveled slate surface will face up.

Another important thing to remember is that there are right-handed and left-handed slater's hammers. You can tell the difference immediately if you hold the hammer in the wrong hand and try to trim a slate—the nail-pulling claw will get in the way.

The stake, a T-shaped piece of steel, supports slate when it's being trimmed. The short leg of the T comes to a point, so the stake can be driven into a rafter. For our roof, though, Hilgert drove the stake into a stump beneath a shade tree and did his cutting there. And to make simple, straight cuts, he often used a non-traditional tool that most slaters consider indispensable today—a tile cutter. The score-and-break technique used for straight tile cuts works fine for slate too.

Nailing it up—Roofing with slate doesn't differ fundamentally from roofing with other kinds of shingles. Overlap from one course to

the next should cover the nail holes of the lower course by at least 3 in., and the joints in one course should be staggered by at least 3 in. from those in adjacent courses. This means that you've got to cut some slates to keep the joints sufficiently staggered. If one course begins at the gable with a full (12-in. wide, in our case) slate, the next course will have to begin with a partial slate. Try not to use partial slates that are extremely narrow (3 in. or less), since these are especially prone to breakage.

Once the framing had been repaired, Hilgert's crew nailed a starter course of slate directly to the wide sheathing along the eaves, overhanging the framing by about 2 in. As shown in the drawing and photos on the facing page, this starter course is laid horizontally, with its length running parallel with the eave. It's best if the starter-course slates are installed face down. This way, the slightly beveled, chipped edge faces downward, creating a better drip edge. After the starter course, all slates should be installed vertically, with their beveled edges facing up.

Proper nailing technique is the most important part of applying a slate roof. If you're used to nailing wood or fiberglass shingles, you'll have to go easy when working with slate for the first time. You're pounding a nailhead that's surrounded by fairly delicate stone, and a single miss can ruin a good slate. A carpenter's hammer can be used, but the narrow head on a slater's hammer is less likely to break the slate surrounding the nail hole.

Nail holes are typically machine-punched at the quarry, but you'll have to hand-punch the slates that are used for hips, valleys and ridges. As shown in the drawings on the facing page, the nail head should sit just below the top surface of the slate. If it's driven too far, the slate around the hole will crack. If it's not driven far enough, the protruding nailhead will crack the slate that overlaps it.

The original slate on our roof had been nailed down with iron cut nails, most of which were in good condition when we removed the slate. Because of this, we decided to use galvanized roofing nails rather than copper nails. Copper is definitely the best choice for slate work because its longevity better matches that of the slate. But copper nails are also a lot more expensive (about $3.00/lb., compared with $.90/lb. for galvanized), so we're hoping that our hot-dipped galvanized nails will last as long as the iron cut nails did. If you want to use copper nails, the sources I've found for

Framing. You can use solid sheathing beneath a slate roof, or roof lath, which was used on the roof shown here. Lath spacing is important, as shown in the drawing opposite, and the eaves and ridge require solid sheathing. Facing page, left: a rotten eave board in the old roof is replaced with new wood.

First courses. The starter course is nailed horizontally to the eave sheathing (facing page, right); then the first vertical course follows. Adjacent slates should be butted together without overlapping. Vertical joints in successive courses should be staggered at least 3 in.

Illustrations: Peter Jennings

Installing slate over roof lath

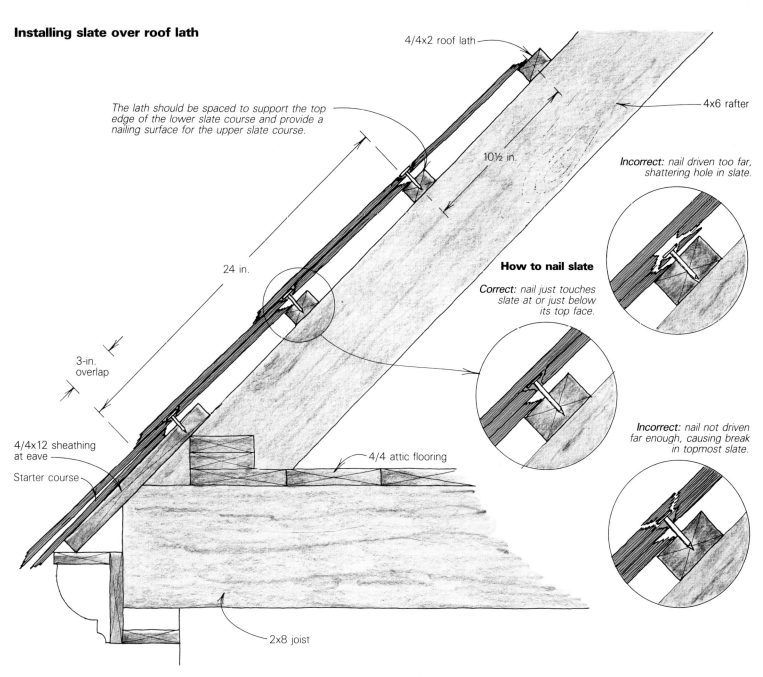

4/4x2 roof lath

4x6 rafter

The lath should be spaced to support the top edge of the lower slate course and provide a nailing surface for the upper slate course.

10½ in.

24 in.

3-in. overlap

4/4x12 sheathing at eave

Starter course

4/4 attic flooring

2x8 joist

Incorrect: *nail driven too far, shattering hole in slate.*

How to nail slate

Correct: *nail just touches slate at or just below its top face.*

Incorrect: *nail not driven far enough, causing break in topmost slate.*

them are Glendenin Bros, Inc. (4309 Erdman Ave., Baltimore, Md. 21213), Prudential Metal Supply Corp. (171 Milton St., East Dedham, Mass. 02026.) and Vermont Structural Slate (Box 98, Fair Haven, Vt. 05743). Correct nail length for standard 3/16-in. slate is 1½ in., for either copper or galvanized nails.

As shown in the drawing on the previous page, each run of roof lath supports the top edge of the slate course below and also serves as the nailer for the following course. To line up successive courses, Hilgert's crew snapped a chalkline down the center of each strip of lath. The upper edge of the next slate course was then laid to this line, leaving about 1 in. of nailing space in the same piece of lath for the following course.

Until the first half-dozen courses of slate had been laid, the crew members could reach the work from the scaffolding under the eaves, or simply by standing on the attic floor. Reaching the higher sections of the roof was more difficult, because they couldn't walk on the installed slates without the risk of breaking them. To reach upper roof sections, they worked from a ladder that they built from 1x4s. The ladder rails are a pair of 1x4s positioned with their broad faces against the roof surface. This provides more even weight distribution than a conventional ladder. A 4x4, cleated across the top of the ladder, holds it in place against the ridges.

Snow guards—Snow and ice accumulation along the eaves can really damage a slate roof. The eave is often the coldest part of the roof, and snow that melts on warmer upper sections can slide down and refreeze at the roof edge. This added weight can cause eave slates to crack and break.

One way to prevent eave icing is to flash the eave with a continuous strip of metal, usually aluminum. The first slate course overlaps the top edge of the flashing by at least 3 in. Only a little snow or ice will stick to the metal before additional accumulations cause the icy mass to slide off and fall to the ground.

The more common approach to eave protection in our area is to use snow guards in above-eave areas of a slate roof. A snow guard (photos above and facing page) is a right-angled metal cleat that is nailed to the lath between slates. Its working edge sticks up above the slates, and is designed to hold snow in place on the roof, minimizing slide-down accumulations along the eaves.

Like our slate, the snow guards we used were recycled. We bought them from another local contractor who had salvaged them when he re-roofed a church in a nearby town. The cast-iron snow guards were at least 75 years old, very rusty, and spotted with roofing tar. We had them cleaned and hot-dip galvanized. All told, they cost us about $6.50 apiece.

On the main roof section shown in these photos, Hilgert installed the guards 4 ft. o. c. in the second and fourth courses. On smaller roofs, you could probably get by with only one row of snow guards. As shown in the photo on the facing page, one slate has to be notched to fit around the snow guard's installation strap. The following slate course then covers this strap.

Flashing and ridge details—As roofs go, the one shown here is simple—no hips, dormers or valleys to contend with, only a couple of chimneys. Because of this, installing the slate was fairly straightforward. But a more complex roof wouldn't be a problem for anyone who's familiar with the hip, valley and flashing details used with wood shingles (see *FHB* #9, pp. 46-50). Chimneys, dormers, skylights and sidewalls that penetrate or intersect with a slate roof should be step-flashed. Hilgert used copper flashing on our roof (with copper nails to avoid any problems with galvanic action), but aluminum, tin, lead and zinc have also been widely used.

Though closed and even round valleys are found on some slate roofs, the open valley is the most common. Install metal valley flashing for a slate roof just as you would for wood shingles. Standards set by the National Slate Association back in 1926 call for open flashing to be slightly wider at the bottom of the valley than at the top to handle the increasing vol-

Snow guards. Nailed to the roof lath between slates, these aluminum or cast-iron elements are designed to hold snow on upper sections of the roof, preventing damaging ice and snow accumulation at the eaves. One slate should be notched to fit around the mounting strap, as shown above. The following course covers most of the strap (facing page). Once you master the nailing technique, a roof with no hips or valleys can be slated quickly.

ume of runoff. Adding about ⅛ in. to valley slates in each succeeding course should create sufficient taper in the open valley.

There are also several options when you come to the ridge. The major ones are shown in the drawing at right. Hilgert finished off our roof with a strip saddle ridge. As the drawing shows, the final full course of slate on one side of the roof extends so that the upper edges of its slates are even with the solid sheathing at the ridge peak. Then these edges are overlapped by the final full course of slate on the opposite side of the roof. The final step is to nail down a second pair of overlapping courses, using partial slates that run lengthwise along the ridge. Nails in this last layer of "combing" slates are positioned so that they fall in the seams between the slates in the course below.

As with a wood shingle or shake roof, detailing at hips can get complicated. Saddle and flashed hips are popular, but you can also use a Boston hip (see p. 24).

Hilgert used no roofing cement to point the seams along the ridge or to cover the exposed nails in final combing courses. Like many slaters, he believes that cement isn't a requirement if a slate roof is installed properly. But in very rainy territory, or if you want to be doubly sure of your roof's weathertightness, use a high-quality silicone caulk to cover exposed nails and to point ridge-course seams. I had some doubts about leakage through the ridge, but the main roof of our house passed its first test for water-tightness the day after it was finished. A driving rainstorm, one of the last in an altogether too-wet spring, hit our part of Pennsylvania. Flashlight in hand, I went up to the attic to look for leaks. I could hear the rain pelting the roof, but not one drop of water found its way through. □

David Heim is a freelance writer based in New York City.

(see p. 24)

Ridge and hip details

The strip saddle ridge relies on overlapping courses along the ridge for weather worthiness. The final courses on each side of the roof overlap along their top edges. They are then covered by two combing courses —slates that run lengthwise along the ridge, overlapping along their top edges and butting along their side edges.

Nail into the seams of previous course and cover exposed nails and holes with silicone caulk.

Combing, or ridge course

Solid 4/4 sheathing at peak

Final full course

4/4x2 lath

Rafter

The saddle ridge, which can also be used for hips, is a more weatherproof design than the strip saddle ridge. The top corners of each slate in the last full courses are trimmed to give nailing clearance for the combing slate. The two combing courses overlap at the ridge, so their nail holes are covered.

Combing slate

Final full course

Rafter

The flashed ridge is well suited to severe weather conditions or to irregular slates that can't be tightly overlapped at the ridge. It's also suitable for hips.

Brass screw through metal strap

Pre-formed ridge flashing

Ridge backing block can be continuous, or spaced every 4 ft. to hold screws.

Rafter

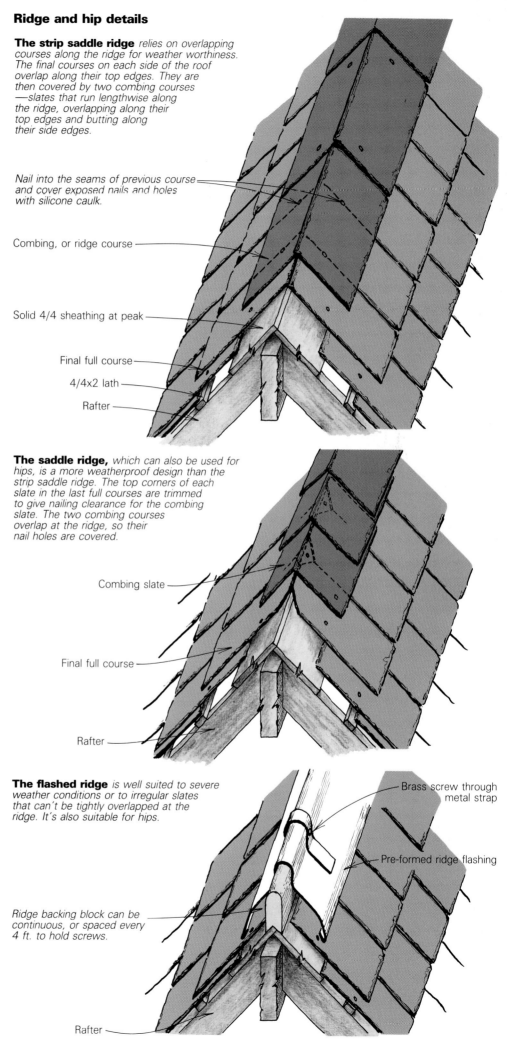

Roofing With Slate

With the right fasteners and flashings, you'll get a roof that can last a hundred years

by Terry A. Smiley

Use good materials for a long-lasting roof. The author uses high-quality materials such as copper flashings and fasteners, bituminous membrane, 30-lb. felt and silicone caulk to ensure his roofs last as long as the slate. Wood strips run up all hips and across the ridge to provide nailing for the saddle caps applied last.

Special tools for roofing with slate. A slate ripper, on the left, is used to cut or pull nails when removing broken slates. On the right are slate hammers. They have a small striking face and a point on the opposite end for punching nail holes.

I worked on my first slate roof when I was 14 years old. The farmer whose house we were remodeling also had a huge barn, and several roof slates were missing. He asked my grandfather if we could replace the slates, and because I was the lightest and the most agile, I was recruited. Ten years later, a customer I was installing a slate floor for asked if I'd be interested in installing a slate roof. By then, I'd worked with a lot of slate and was ready to give a whole roof a try. I called the supplier from whom I'd gotten flooring slate, and he sent me a copy of *Slate Roofs*, first published by the now-defunct National Slate Association in 1926. Everything worked out fine, and since then I've put on many slate roofs from Pennsylvania to Arizona. (*Slate Roofs* has been reprinted by Vermont Structural Slate Co. Inc., P. O. Box 98, Fair Haven, Vt. 05743; 800-343-1900. The book sells for $11.95.)

Slate is the ideal roofing material. It won't corrode or burn, and for the most part it won't wear out. I'm not saying that a slate roof will never need replacement or repair. But it is important to keep in mind that a roof is a system of its materials, and a slate roof is made up of more than just pieces of sedimentary rock pulled from a quarry, split by hand and cut into rectangles. The integrity of a slate-roof system comes from all its components—the slate itself, the flashings, the fasteners and the underlayment—acting in concert, and the failure of any component can result in the failure of the whole system and leaks in the living room.

A slate roof is expensive; slate can cost between $250 and $600 per square (100 sq. ft.), and depending on the complexity of the roof, my labor costs can run as high as $450 per square. Given the high cost of the slate, it's foolish to scrimp on the quality of the other components. Using second-rate materials for any part of a slate-roof system is akin to running recapped tires on a Rolls-Royce.

I use 16-oz. copper for all flashings, copper nails, bituminous membrane at all the eaves, silicone caulk and 30-lb. felt or roll roofing under the slate. Although some people use standard, galvanized drip edge, a metalsmith makes my drip edge for me out of 16-oz. copper. I have a standard drip-edge profile, which I adjust to fit the house and roof design. It covers at least 1½ in. of the fascia and runs up the roof a minimum of 4 in. I run it along the eaves and up all the rakes. I bend my own step flashings, but there isn't any reason why your tin knocker couldn't make these pieces for you as well.

Order your slate well ahead of time—Unless you live near a slate quarry, it's unlikely that you'll be able to go the local roofing supplier and pick up enough slate to do a roof. My regular supplier—I've been using New England Slate for 10 years (Burr Pond Road, Sudbury, Vt. 05733; 802-247-8809)—will send me a slate-availability list (along with samples) to match the specification of the job. The list includes an approximate delivery time as well as estimates of delivery costs, via tractor-trailer. My typical slate order takes anywhere from one month to three months to arrive at the job site.

Pipe staging holds a lot of weight—Because slate is so heavy, loading it on a roof can be a precarious operation. And it seems that no matter how carefully I try to plan things, if I load the roof with more than a couple of days' worth of slate, I spend a lot of time moving the darn stuff around. So I prefer not to stock the whole roof at once. I think it's safer to keep roof surfaces open and free of clutter. And it makes layout easier if the roof isn't full of roofing materials.

It goes without saying that slate is heavy. You need a strong staging on which to stock materials. I like to use steel-pipe scaffolding. It's quick and easy to set up, and although you can rent pipe staging, three months of rental fees about equal the purchase price. If possible, I have a scaffold at all the eaves. I also wear a safety harness when I'm working on a roof (sidebar p. 56).

Thirty-lb. felt and bituminous membrane dry-in the roof—Putting on a slate roof is slow (but satisfying) work. A typical job takes me about three months after the time the slate arrives from the supplier. One huge job I did—170 squares, laid in a graduated, textural pattern—took me a year to complete. Because so much time passes from start to finish and because it's likely that the other trades are finishing off the interior of the house while I'm working, I take special care to dry-in the roof.

Roof dry-in starts with the application of Ice and Water Shield at all eaves (W. R. Grace & Co., P. O. Box 620009, Atlanta, Ga. 30362; 404-448-5880). This bituminous membrane is designed to stop water from backing up under the slate with the formation of ice dams, a problem in northern climates for all roofing materials, not just slate. Ice and Water Shield is a self-adhesive flexible membrane, 40mm thick. It's made of polyethylene film and rubberized asphalt. On the back it has a release paper that is peeled off when the membrane is applied to the roof deck.

I lay a 1-ft. wide strip of membrane on the roof deck starting at the fascia. Over this first piece I

A slate hammer pokes a small hole in the back of a slate and makes a beveled hole in the face. A sharp blow with the point of the slate hammer punches a small hole in the back of the slate about 2 in. from the outside edge. The hammer's point pokes a small hole (the size of the hammer point) through the first couple of layers of the slate, and the percussion of the hit blows out the rest of the layers, leaving a conical hole. Ideally, the larger hole on the slate's face makes a countersink for the nail head.

install the copper drip edge. The top 3 in. of the drip edge then are covered by a 3-ft. wide sheet of membrane. Another sheet goes above this one, lapping the first 3-ft. sheet by 3 in. Two 3-ft. sheets of membrane give me almost 6 ft. of protection from ice damming.

After the second row of bituminous membrane is on the roof, I complete the dry-in with rows of roll roofing or of 30-lb. felt. Fifteen-lb. felt is the standard weight for most roofing materials' underlayment. But the additional cost of the heavier materials I use is offset by the fact that they hold up longer without repair.

Roofing nails will hold the paper securely, but I like to use special nails called cap, or button, nails. These nails have a small square of stiff plastic pushed onto the shaft right below the head. The large head of the cap nail holds the larger piece of plastic securely against the felt, effectively resisting the forces of wind and rain.

Two special tools are all you'll need to get started—For around $100 you can get all the special tools needed to put on a slate roof: a slate hammer and a slate cutter. Both are available from New England Slate.

A slate hammer is essential to good slate work. Most slate hammers are lightweight, about 14 oz., and they have a small striking face, usually ¾ in. across. Slate hammers have a long, tapered 6-in. point on the back of the head (right photo, p. 52). The hammers' point is used for punching nail holes in slate.

Most slate will come from the supplier with a hole punched in each top corner, but there are a lot of times, such as when you cut a large piece of slate into smaller pieces, when you'll have to punch your own holes. It's not hard to do, but like a lot of simple procedures, punching holes takes a little practice.

Because I'm right-handed, I hold the piece of slate face down in my left hand and then give the back of the slate a sharp blow with the point of the slate hammer. The face, or the exposed side of the slate, is the side with the beveled edges. When the slate is cut at the quarry, the shearing action of the cutter leaves a beveled edge on one side and a smooth edge on the back. I try to punch the hole about 2 in. from the side edges of the slate and 1 in. more than the exposure line from the bottom. Just the right hit will poke a small hole in the back of the slate (top photo, p. 53).

When you turn the piece over, you'll see that the percussion of the blow has blown out a larger hole on the slate's face. The point of the hammer pokes a small hole (the size of the hammer point) through the first couple of layers of the slate (bottom left photo, p. 53), and the percussion of the hit blows out the rest of the layers, leaving a conical hole. Ideally, the larger hole on the slate's face makes a countersink for the nail head (bottom right photo, p. 53).

For small pieces of slate, say the little pieces of a hip's cap, which a hammer blow might break, I lay the slate face down on a board and punch the hole with the hammer's striking face and a nail set. The less violent blow to the slate sometimes can keep it from breaking.

A slate cutter looks like a paper cutter. The author holds the finish-side down and uses short, chopping strokes to munch through the piece of slate. The cutter gives a clean cut on the back of the slate and a beveled cut on the face or finish side.

The first course is made of three layers of slate. The first and second layers of slate, cut to 4 in. and 10½ in. respectively, are bedded with silicone and nailed even with the drip edge. The third layer is full-length slates. For a hip roof, the full-length slate is notched around the wood strips that provide nailing for the saddle caps.

Layout tools. The author uses a wax pencil for laying out courses on a sheet-metal story pole. The red wax shows up equally well on slate, sheet metal and felt paper.

Several other features also are incorporated into most slate hammers. These features include a nail puller and a slate-cutting edge that is forged into the metal shaft between the handle and the hammerhead.

Some people use the hammer's slate-cutting edge for trimming pieces of slate. I prefer to use a slate cutter, which is similar in design to a paper cutter. The cutting edge is anywhere from 4 in. to 16 in. long.

Slate cutting is a process of shearing, or nibbling, through the slate, rather than a guillotine action. You put the slate in the cutter face-side down. You hold the slate with one hand, and with the other hand you force the cutting edge through the slate in short strokes to nibble away at the stone (photo top left). Slate cutting is surprisingly easy, and with practice any intricate shape can be cut.

The slate cutter forces the layers of slate away from the exposed face, and when the stone is turned over, the cut edge is beveled away from the cut. Occasionally, if I am using thick slate or if I am working on a special detail where a square, nonbeveled edge is required, I'll use a wet saw with a diamond blade to make my cuts. For the most part, however, a slate cutter is faster, if not more convenient.

Sheet-metal story poles speed layout— Roofing slates are available in different lengths, and the slates' length establishes the weather exposure of the courses on the roof. Longer slates are laid with longer courses. The slates used for the photos in this article are 18 in. long. Other standard lengths are 12 in. and 24 in.

Slate is laid on a roof so that a part of each slates' length is covered by the next two courses above. Because of this overlap, the roof is always three slate layers thick. Slate is laid with a 3-in. head lap. This construction means that the top of each course is covered not only by the next course above but also by the first 3 in. of the second course above. The 3-in. head-lap rule is the basis for figuring the course height of a certain length of slate. To figure out the course height, you take the total length of the slate, subtract 3 in. and divide the difference by 2.

For instance, an 18-in. slate minus 3 in. equals 15 in. Dividing that in half leaves a 7½-in. course for an 18-in. slate. Course height can vary by ¾ in. short or ¼ in. long. In order to run an even course up to a ridge or to ensure a course meets, say, the bottom of a skylight, you could run 18-in. slates in courses from 6¾ in. to 7¾ in.

I like to lay out all my slate courses for the whole roof before I nail on the first piece. Establishing my course layout ahead of time prevents confusion when I start laying slate. I lay out my courses on story poles made of 24-ga. galvanized sheet metal that I rip to 2 in. (photo bottom left). I pop rivet the strips together in lengths equal to the eaves-to-ridge length of the roof.

To make a matching set of story poles (one for each end of the roof), I nail two strips next to each other at the ridge, letting them hang down the roof to the eaves. With the two strips hanging next to each other, I can mark my course lines on both strips at once. I like to use a mechanical

Duct tape and string help to align the saddle caps. The pieces of the ridge cap or hip cap (shown here) are bedded in silicone and fastened with two nails each into the wood strips the author attaches to all hips and ridges. To keep the caps straight, he sights along a string that runs from the eaves to the peak. Strips of duct tape hold the two pieces of the caps together while the silicone dries.

red-wax pencil for marking the story poles (bottom photo, facing page). I also use the wax pencil to mark on slate.

Once I have the courses set up on the story poles, I nail one at each edge of the roof and then snap lines, usually six courses at a time. I then roll the story poles to the ridge, where they are out of the way. I first used this system on a complex roof that had seven different slate lengths and different course heights, and since then, I realized that it saves time on even the simplest layout.

Slate nails act as hangers—Slate nails are a critical part of a slate roof's life expectancy. The best nail for slate work, combining long life and a wide range of sizes, is copper. Standard copper slating nails are available in sizes from ½ in. to 3 in. To figure nail length, double the thickness of the slate and add ¾ in. for deck penetration. For this roof I used ⅜-in. slate and 1½ in. nails

Nailing is the most important skill required for good slate work. A nail must be driven far enough below the surface of the slate (into the countersunk hole made by the hammer) so that the slate above won't rest on the head of the nail and provide a stress-inducing high spot. Conversely, the nail must not pull down so hard on the slate as to break it. The best way to nail a slate is to hold it down snug with one hand and then sneak the nails in just below the surface. Experience will teach you when you've nailed wrong, but only at a price.

Slate nails act more like hangers than fasteners. They are holding the slates on the roof, not holding them to it. If you could pick up a proper-

A capped hip is six layers thick at the eaves. A three-layer first course is capped by a three-layer saddle cap for a total of six layers of slate.

ly nailed slate roof and give it a good shake, the slates would be loose enough to rattle.

Triple-layer first course is bedded in silicone—In roofing with slate, the first course, starting at the eaves, is three layers thick (center photo, p. 54). I cut the slate of the first layer to a length of 4 in. I punch holes in these small slates and nail them even with the bottom of the copper drip edge. Before nailing, I bed the slates in two walnut-size blobs of clear silicone caulk. I bed all my small slates—starter courses, ridge caps, small pieces running up a hip—in silicone.

The silicone provides long-term shock protection against things such as someone leaning a ladder against the eaves or walking up a hip. The silicone also provides additional resistance to the unsightly slippage that can sometimes happen with smaller pieces of slate. Few things look worse from the ground than a slate that has slipped below the others in an otherwise straight course on a roof.

The length of the second layer in the first course is determined by adding 3 in. to the full-size course height. If I'm using 18-in. slate and if the exposure of the courses is 7½ in., then the second layer is 10½ in. long.

The second layer is also bedded in silicone, and it is laid over the 4-in. first layer, even with the drip edge. Next comes my first full-length layer of slate. It covers the first two layers of the starter course and is laid with its bottom edge even with the drip edge. From here, standard coursing, following the lines I've snapped with my story pole, continues to the ridge.

Along with a 3-in. head lap, I lay slate with at least a 3-in. side lap. This layout means that joints between slates in one course should be offset by at least 3 in. from the joints in the course below.

Use wood strips for the ridge and hip saddles—I cap all ridges and hips with slate saddles. Each saddle is made of two pieces of 12-in. by 6-in. slate that butt together and extend 6 in. down each side of a ridge or hip. I use a 3-in. head lap for the saddle. Following the same formula used to determine course exposure for the full-length slates gives me a saddle exposure of 4½ in. (12-3 =9; 9/2=4½.)

As a base for the saddles, I rip pieces of wood 2 in. wide and as thick as the three overlapping layers of slate that result from the slate course running up the roof. The wood provides a solid base to nail the slate saddle caps to. I nail the pieces of wood along all of the ridges and down all of the hips (left photo, p. 52), stopping 6 in. from the eave end of the hips and 6 in. from the rakes or gable. If the saddle base pieces went all the way to the eaves, they would be visible from the ground.

Running slate up a hip roof is not difficult, but it does require some careful slate cutting. I make a three-layer first course similar to the straight-eaves first course described earlier. The only real difference is that the pieces have a 45° cut along the edge of the hip and a notch cut in the second and third pieces where they go around the wood saddle base pieces (center photo, p. 54). After the starter course is run, finishing the hip is a mat-

ter of filling in pieces of slate where they meet the solid-wood strips.

At the eaves, a hip is six layers thick—For a hip roof, I start laying the saddle caps at the eaves. Like all first courses, the first course of saddle caps has three layers. The three-layer hip and the three-layer saddle make a total of six layers

Fall Prevention

I think of scaffolding as a way to get materials to the roof, not necessarily as something that keeps me from falling off. To keep from falling, I rely on gear similar to what mountain climbers use (photo above). I wear a harness that has a sling around each leg and a belt around my waist. A carabiner, which is a big clip kind of like a giant safety pin, attaches to both leg slings and to the belt. I run a series of screw eyes into a rafter every 12 ft. along the roof ridge. A climbing rope is tied to another carabiner and hooked to the screw eye. A smaller rope runs from my harness to the climbing rope. A simple slipknot, the barrel knot, connects the smaller rope to the climbing rope. The barrel knot allows me to move up and down and across the roof while keeping a taut line between me and the ridge.

You'd be surprised how fast you'll get used to wearing the harness and using the ropes. For me the climbing gear is much more comfortable and infinitely more adjustable than other construction safety belts I have seen. Clerks at a mountaineering store can set you up with everything you need, and they can show you how to use the equipment.

I've been a firm advocate of a rope and climbing harness ever since I was on a job where a man fell 8 ft., a man who is now a paraplegic.—*T. S.*

of slate at the eave end of the hip (bottom photo, p. 55). I punch two nail holes in the upper left-hand or right-hand corners of each saddle piece, depending on which side of the hip they will go on. The holes have to be close together so that the nails will go into the 2-in. wood strip instead of through the slates below (top photo, p. 55).

As I run the saddle caps up a hip or across a ridge, I bed each piece of cap in silicone. I run a bead of caulk between the two pieces of slate where they meet at the ridge. I stop the bead 4½ in. from the bottom edge so that the caulking won't be visible. Triple coverage of the slates keeps things watertight. Even though every cap is held in place by the two nails in each piece, I use duct tape to hold the saddle slates in place until the silicone dries (top photo, p. 55). I run a string down the center of a hip or ridge. Aligning the joint where the pieces of each saddle cap meet with the string gives me a straight course.

A slate ripper removes broken slates—I always give my customers a limited lifetime guarantee on their slate roofs, but all they have to do to void the guarantee is walk on the roof. Slate is brittle, and it will break easily. If a slate roof is never walked on, the most maintenance you may need to do is replace a couple of slates that inevitably get broken when the roof is installed. When I'm installing a slate roof, I always walk on it gingerly, watching where I step.

Replacing a slate is accomplished using a special tool, a slate ripper. A slate ripper is an 18-in. to 30-in. long piece of flat ⅛-in. by 2-in. forged steel attached to a round, offset handle (right photo, p. 52). A hook is forged on either side of the flat end of the ripper.

To replace a slate, I slide the ripper under the broken slate and use the flat end to locate and then hook over the nails holding the slate in place. Once the nail is hooked, I smack the offset handle of the ripper with a sharp blow from a 2-lb. sledgehammer. A single sharp blow causes less stress to the surrounding slates than repeated light blows.

Once the nails are pulled or cut, the pieces of the broken slate should slide out. A new slate of similar size is bedded in silicone and slid into place. To fasten the new slate, I snip both sides of a nail head to make it T-shaped. I punch a hole in the replaced slate at the upper end of the slot between the slates in the course above. Then I drive the T-headed nail between the slot and through the slate.

Next I fashion a piece of copper to cover the nail in the slot between the slates above. The copper is 3 in. wide and long enough to hook over the top of the replaced slate and cover the T-nail by 3 in. I hem, or bend over, the top ¾ in. of the copper. Then I slip the copper, hemmed-side down, onto the flat end of the ripper and shove the ripper between the replaced slate and the slates above. The hemmed end should snap down past the top of the replaced slate and hook onto the edge. Like all slate-roofing techniques, all this one takes is a little practice. ☐

Terry A. Smiley runs TAS Construction in Woodland Park, Colo. Photos by Jefferson Kolle.

Slate Quarrying and Shingle Manufacture

Heavy equipment gets the slate out of the quarry, but shingles are still split by hand

by Jeffrey Levine

Throughout America's slate-quarrying regions, roofing shingles are still split by hand with hammer and chisel, just as they were more than 150 years ago. Indeed, it is the rough, hand-wrought look of slate shingles that is largely responsible for slate's continuing popularity as a roofing material.

The labor-intensive nature of slate-shingle production has a price, however—as much as $475/sq. for premium Virginia slate. But a slate roof, the lifespan of which can be as much as 150 years, still represents an exceptional long-term value. In this article I'll take a look at slate as a raw material and at its conversion into roofing shingles. Understanding this process will contribute to your appreciation of this traditional, durable product.

Formation and structure—Slate is a metamorphic rock that originates from sedimentary particles of clay and fine silt. These were carried by streams and deposited on the bottom of ancient seas, forming successive parallel and horizontal layers, or beds. The pressure of subsequent layers of material eventually consolidated this clayey silt into shale, which is a general term for laminated rock composed of clay and mineral carbonates. Geological forces exerted even greater pressure and heat on the shale, chemically changing the original clays into new minerals, including mica, quartz and chlorite, and transforming the shale into slate.

Slate possesses two significant structural characteristics: cleavage and grain (drawing, p. 60). Cleavage, a result of the parallel layering of the mineral flakes that comprise slate, is the property that most differentiates slate from shale and the other sedimentary rocks and that allows slate to be split along parallel planes. Whereas shale may be split along its original bedding planes (the layers of original sediment deposition) into thin sheets, slate can be split at any point in its cleavage plane into thin, tough sheets.

Grain is the second natural plane along which slate may be split. It is much less pronounced and usually runs at right angles to the direction of the cleavage. Grain can sometimes be seen as striations on the cleavage surface, though it ranges from being quite distinct to being nearly invisible.

From bed to block—Until the 1870s, the quarrying of slate was a rudimentary affair. At that time, the simple tools used to quarry stone—picks, wedges, crowbars, gunpowder, and horse-powered windlasses—gave way in part to more advanced equipment such as the steam stone channeling machine, the steam

The quarry. This 250-ft. deep quarry, owned by Anthony Dally and Sons of Pen Argyl, Pa., is typical of slate quarries in northeastern Pennsylvania. The diamond belt saw in the left foreground can also be seen in the photo on the next page.

Three generations of stone-extracting technology. Prior to the Industrial Revolution, quarrying tools consisted of hammers, chisels, prybars and explosives. With the advent of steam engines, the steam stone channeler was developed (photo above left); it dramatically reduced the waste associated with explosives. Subsequent advances in quarrying technology include the wire saw (photo above right), first used in this country in 1926, and the diamond belt saw (photo below), introduced in the late 1980s. ***Drawing and photo above courtesy of Jeffrey Levine.***

hoist, and the mechanical drill. These new devices greatly increased productivity, while substantially reducing waste.

The steam stone channeling machine was mounted on railroad track in the quarry, and looked like a locomotive engine with a giant side-mounted reciprocating chisel (top left photo). The channeler cut narrow vertical grooves in the rock, thereby freeing up the sides of large blocks of stone. After channeling, the block was undercut with wedges in order to split the block from its bed. The channeler made possible the raising of larger and more uniform blocks, with less labor and waste of stock than was possible with blasting.

The next significant advance in U. S. slate-quarrying technology didn't come about until 1926, when the U. S. Bureau of Mines conducted trial tests of a late 19th-century Belgian invention. The wire saw (which is still in use in many quarries) consists of a three-strand steel cable, roughly ¼ inch in diameter, that runs as an endless belt (top right photo). A drive-pulley powered by an electric motor and a tension-pulley are located outside the quarry. The equipment in the quarry consists of a pair of standards set up 60 to 100 feet apart, each with a fixed upper and movable lower sheave for receiving and conducting the wire. The standards are placed in holes about 10 ft.

Monster wetsaw. Essentially a huge radial-arm saw, this saw makes quick work of crosscutting the large blocks of slate into more manageable pieces. The 36-in. blades cost about $1,400 apiece and last eight to ten weeks in daily use. The saw is operated by an operator who sits in a control box located at the end of the travel arm.

Sculping. Setting his chisel vertically on the end-grain of the previously crosscut block of slate, this worker sculps the slate into the blocks from which the shingles will be split. Fewer than a half-dozen blows are usually sufficient to break the slate. The sculped block is then pried away from the larger block and rolled down to the splitting area.

Splitting shingles. With skill and precision this workman splits shingles in the same manner as his great-great-grandfather might have (left photo). Using hammer and chisel he fractures the slate, then gently pries the shingle off the remaining block if necessary. A stack of "eights," or blocks from which eight shingles will be split, is to his left. In the middle photo, he "favors" the slate, causing the quartered section of an eight to separate into two shingles.

Trimming. With quick, dextrous motions, this man trims each side of a shingle to bring it to a uniform dimension. The process takes just seconds. The trimmer above resembles an over-size paper cutter. *Photo by Jeffrey Levine.*

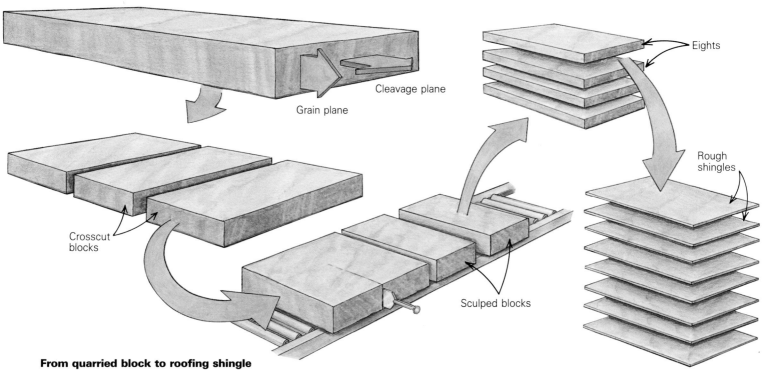

Eights

Cleavage plane

Grain plane

Rough shingles

Crosscut blocks

Sculped blocks

From quarried block to roofing shingle

When the blocks of slate are brought to the splitting shed, they're first crosscut with a diamond wetsaw. The crosscut blocks are then sculped, or split along the grain, with hammer and chisel. These blocks are split, often into smaller blocks called "eights," which are then halved again and again to produce eight shingles apiece. The rough shingles are mechanically trimmed and punched to produce the finished product.

deep and wide enough to accommodate the movable sheaves (the holes themselves are made with core drills). By lowering the sheaves, the wire is brought into contact with the slate and, when fed with sand and water, makes a cut over the entire distance between the standards.

The wire saw further reduced the proportion of waste and cost of production of slate. It was most widely adopted in the Pennsylvania slate districts of Northampton and Lehigh. Blasting techniques continued to be used in the other slate districts of the U. S., and in fact are still used in some instances.

The most recent innovation in slate quarry-

ing is the introduction of the diamond belt saw to U.S. quarries (photos, pp. 57 and 58). The technology has existed in Europe for nearly a decade, but has only recently found its way to this country. Similar in principle and in operation to the steam stone channeler, these saws look and function like giant chainsaws. They are operated from a control panel on a boom that pivots 360° around the housing. The blade is let into the stone gradually, then the whole unit moves down a track at about 1½ in. per minute. At $6,000 apiece and with a lifespan of only four to six weeks, belts for this saw still represent a savings for the quarry owner. Even with this latest advance, however,

wedges and sledge hammers are still used to separate the channeled blocks from the bed.

From block to shingle—Manufacturing slate roof shingles is essentially a five-step process: cutting (across the grain), sculping (splitting the stone along the grain), splitting (in the cleavage plane), trimming, and hole punching. Though certain steps in the process have been mechanized, no machine has yet replaced the human hand in the sculping and splitting of slate.

Large blocks are taken from the quarry to splitting sheds via aerial cableways, or by loading the blocks on flat-bed trucks. Blocks gen-

Drawings above: Bob Goodfellow

Slating tools, and a primer on replacing broken slate shingles

There are three tools both unique and essential to the traditional slater's toolbox: the slate hammer, the ripper and the stake (photo right). A fourth tool, the slate trimmer, has largely replaced the stake and hammer for trimming shingles in recent years. All are available from John Stortz & Son, Inc. (210 Vine St., Philadelphia, Pa. 19106, 215-627-3855), or you might find them through an antique dealer or at a tag sale. Prices for antique tools, oddly enough, are usually lower than for new tools.

The slate hammer is a multipurpose tool with a small, round or oval-shaped head for hammering nails, a point opposite the head for punching holes and a knife edge on its shaft for trimming and cutting. At the juncture of head and point, on the side of the hammer, is a claw for removing old roofing nails. The handle is often wrapped with leather rings in order to provide a good grip and to prevent blistering of the slater's hands.

The ripper is a thin piece of iron or steel, measuring between 18 in. and 30 in. in length, with an offset handle at one end and a pair of hooks at the other. The thin blade is slid under slates to be removed, hooked onto the slating nail, and then hammered back out with a claw hammer (photo below left). When the nails have been removed in this way, the slate will slide out.

The stake is a T-shaped piece of iron, about ¼ in. thick and 18 in. long, upon which the slater places the edge of a slate to be trimmed. The short stem tapers to a point so that it can be driven into a scaffolding plank or rafter.

The first step in replacing a broken slate is to remove it, using the ripper. If the slate does not slide out by itself, the pointed end of the slate hammer can be punched into the slate through the joint between the slates above, and the slate dragged out. The replacement slate, which should match the old slate in color, texture, shape and color permanence, is then slid into place. The new slate is held in position by one nail inserted through the vertical joint between the slates in the course above. A piece of copper flashing, roughly 3 in. by 8 in., is then slid lengthwise under the joint between the two slates above the new slate, and over the nail, until it extends an inch or so under the tail of the slate two courses above the replaced shingle. The copper "bib," as it's called, should be bent in a concave shape and teeth formed along its longer edges before insertion to ensure that it will remain in place (photo below right). —*J. L.*

erally measure from 4 ft. to 6 ft. in length, from 2 ft. to 4 ft. in width, and from 4 in. to 6 in. in thickness.

The blocks are first crosscut with a large diamond wetsaw (top photo, p. 59). Sections, ranging from 12 in. to 26 in., are cut so as to maximize the amount of usable material and to produce blocks slightly longer than the length of the finished roofing slate. Prior to the advent of the diamond wetsaw, blocks taken from the quarry were broken across the grain with chisel, mallet, sledge and wedges.

Once crosscut, the smaller blocks are either slid or hoisted to the sculping area where they are sculped to widths slightly larger than the width of the finished slate (bottom photo, p. 59). Sculping is still accomplished with a mallet and a broad-faced chisel. Roofing slates are always fabricated with their long sides in the direction of the grain in order to minimize the possibility of breakage when installed on the roof.

The sculped blocks are next moved to the splitting area where they are split along their cleavage planes to thicknesses ranging from 3/16 in. for commercial standard slate to 2 in. for architectural grades. The blocks are generally stacked side by side, on edge, adjacent to the splitter's workstation. The splitter keeps them moist prior to splitting by periodically wetting their edges with a rag or sponge, and by sometimes keeping the blocks covered with burlap and a layer of mud. This is done because slate that has been allowed to dry out does not split well.

The splitter places the block along his thigh. Swinging a wooden mallet or small sledge hammer against a thin, flexible stone-splitting chisel, he first creates a fissure in the block (top left photo, facing page). Then he "favors" the slate. This is done by prying gently with the chisel once it has entered the block, allowing the slate's natural springiness to extend the split. The process of halving the split portions is repeated until shingles of appropriate thickness are obtained.

The rough shingles are then trimmed using electric or treadle-powered machines (top right photo, facing page). These use either a revolving or pivoting blade into which each of the four edges of the slate tile is thrust. The slates are supported as they're being cut by metal plates attached to the machine. These supports are also marked to permit the operator to gauge the proper shingle dimensions.

The final step in slate-shingle manufacture is to punch two nail holes into the tiles. Nowadays, this is done with a compressed-air or treadle-powered punching machine, which punches both holes simultaneously. But in the days before the advent of mechanical tools, a shingle maker would have trimmed each slate by hand and punched two holes in it with the point of his shingle hammer. □

Jeffrey Levine is an architectural conservator with John Milner Associates, West Chester, Pennsylvania. Photos by Vincent Laurence, except where noted.

Removing a damaged shingle. The author hammers on the ripper to remove the nails holding the shingle in place.

Finishing a slate repair. The concave form and toothed edges of this copper "bib" help to ensure a good friction fit.

Photos this page: Jeffrey Levine

Tile Roofing

Clay or concrete, a tile roof will outlast the next generation

by J. Azevedo

In an 1895 technical book on tilemaking, author Charles Davis listed as a disadvantage of tile roofing that it is "anything but handsome." He was not the first to make disparaging remarks about the appearance of tile. Its popularity has cyclically risen and fallen in this country for the past 300 years. The recent resurgence of tile in the residential American roofscape—along with the consequent assumption that many homeowners must find tile to be quite handsome after all—reminds us that tastes do change. And in the case of tile's return to favor, changing taste coincides with good sense—the advantages of tile roofing far outweigh its disadvantages.

All tiles worth considering come with a 50-year guarantee, a measure of their durability in all climates. In wet climates, tile will not rot. Moss and fungus cannot get a foothold, bugs find it unappetizing, and rodents and squirrels chip their teeth on it. In cold climates, although tile will not shed snow as a metal roof will, well-made tile will withstand repeated freezing and thawing (most Swedish houses are roofed with tile). In coastal areas or cities, tiles can be used with impunity because they are unaffected by salt spray or pollution. In hot, dry areas, tiles will not split or crack. Also, tile roofing is available in light colors to reflect the sun's rays. A tile roof will not stop a fire that starts in your kitchen, but it will keep a fire in your neighbor's kitchen from spreading to your house through his roof. With its Class A fire rating, tile has the approval of every fire code from Boston to Los Angeles.

Finally, tile roofing in its many forms has been around so many years that its use in a new house can recall a certain region or historical period. Some tile profiles are synonymous with California missions or French country estates or oriental temples. Yet, not every tile has a singular connotation. A modern concrete tile with a profile loosely based, for example, on a Mediterranean tradition can look quite at home on a house in the Northwest. In fact, the array of tile profiles and colors available today give tile the virtue of versatility. Tile roofing can ripple lightly across the roof or sit elegantly in solid silence; it can form neat symmetrical vertical ribs or take playful dips and leaps. If only Charles Davis could see them now.

It began with clay—Clay-tile roofing originated in neolithic China and later, independently, in the Middle East. From these two loci, tile roofing spread throughout the Orient and Europe. In China, tile roofing has become vernacular. The Japanese adopted tile roofing, adapted it to their harsh climate, and now

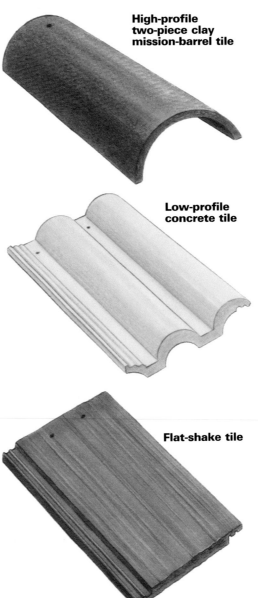

High-profile two-piece clay mission-barrel tile

Low-profile concrete tile

Flat-shake tile

have a major tile industry, including the largest clay-tile factory in the world. The ancient Greeks and Romans roofed their buildings with tiles of clay or, to show off, marble, and the Europeans picked up the clay-tile idiom and carried it to the present.

European settlers brought the tradition of tile roofing to America, starting at both coasts and working towards the middle. On the East Coast, Dutch settlers first imported tiles from their homeland but by 1650 were making them here. In the West, Spanish missionaries brought tile roofing along with their religion.

In California, tile folklore recalls how Indians made the first mission tiles from riverbank clay, formed across the curve of their thighs. Modern manufacturing methods, less sensual but more precisely controlled, generally start with shale, which is crushed to a fine powdery clay. The clay is mixed with water and kneaded (pugged) to the consistency of cookie dough.

To form simple clay tiles, moist, plastic clay is extruded through a die, like dough through a cookie press, and sliced into lengths. More complicated shapes come from pressing the clay into molds. Some ornamental tiles are even sculpted by hand. The formed tiles dry in a room kept at the temperature of a warm summer afternoon, and from there they go to a kiln for firing. For their trip through the kiln, the tiles are either stacked on a refractory (like a small railroad car) that is pulled through a tunnel kiln, or laid on ceramic rollers that convey individual tiles through a roller-hearth kiln. Regardless of the method, the object is to raise the temperature of the clay to the point of vitrification (about 2000° F), at which point the clay minerals lose their individual identity and fuse.

Clay tiles are available in colors to suit any taste, from subtle to outlandish. Right out of the kiln, the tile has the natural earth-tone of the original clay. A nearly pure clay will fire to buff, while iron-oxide impurities impart the terra-cotta color typically associated with Spanish tiles. Manufacturers can also mix clays to get a range of natural colors.

To make variegated tiles, a manufacturer can inject a burst of gas into the kiln, giving the tile a random scorched streak called "flash." Another way to color tiles is to spray a

Drawings: Bob LaPointe

High-profile tiles are typically made of clay, in either the traditional two-piece mission-barrel style, or the one-piece "S" tile shown here. To discourage nesting, this roof has clay plugs, known as "birdstops," under the courses of tile adjacent to the eaves.

Low-profile tiles are designed to interlock at the edges to keep out the weather. These variegated concrete tiles have been colored with a slurry of cement and iron-oxide pigments.

Flat tiles can be colored and textured to resemble traditional roofing materials such as slates, or the wooden shakes suggested by the dark, striated patterns of the concrete tiles on this roof.

thin creamy layer of clay (called a slip) onto the tile before firing. The tile will then take on the color of the slip. The most dramatic, and most expensive, way to color tile is with a glaze (photos, p. 67). The metallic pigments in the glaze, when fired, melt to a glossy, vitreous, richly colored surface, much like that found on ceramic tile used indoors. Manufacturers who offer glazed roof tiles stock a range of deep brilliant colors and can also brew custom colors.

The long history of clay-tile roofing has produced a variety of styles (drawing, p. 62). The industry typically divides them into three groups: high profile, low profile, and flat. Today you'll find examples made of clay or concrete in all three categories.

Currently, the most popular style is the high-profile "mission" (top photo, previous page). It comes in either the traditional "two-piece barrel" (the concave trough—or *tegula*—and the convex cap—*imbrex*—which is a trough tile turned upside down) or the newer "one-piece S," which joins the trough and cap into a single tile.

Low-profile tiles have a more precise, interlocking look (photo below). Most low-profile tiles are made of concrete, and are characterized by linear bevels, arches and grooves. They are made to interlock along their edges (drawing, facing page) to keep out wind-driven rain and snow.

Flat (or shingle) tiles are just that. They range from simple unglazed rectangles of clay or concrete to textured and glazed flat tiles meant to resemble the variable topography of slates or the coarse pinstripes of thick wooden shakes (bottom right photo, previous page). Like the low-profile tiles, flat ones typically have interlocking edges to keep out the weather.

A few manufacturers can do custom work to match an existing tile that has gone out of production, or to produce decorative finials or other roof ornaments. Ludowici-Celadon, Inc.

(P. O. Box 69, New Lexington, Ohio, 43764) and Vande Hey Raleigh (1665 Bohm Dr., Little Chute, Wis. 54140) are two such companies.

Concrete-tile roofing—Compared with clay, concrete tile is the new kid, with a history of just over 150 years. A German farmer is credited with making the first concrete tiles in the early 1800s for his barn. Commercial production started in Bavaria in the mid-19th century. When concrete-tile roofing spread to the rest of Europe around the turn of the century, manufacturers began to add color to the tiles. Still, acceptance of the new product was slow to come. A 1930's book on tiling noted that concrete tile, "…generally considered a new industry, is subject to a good deal of prejudice," which the industry countered with a vigorous public relations campaign. One company even named itself "The Everlasting Tile Co." Europeans no longer need to be convinced—90% of European houses are roofed with concrete tile. The same holds true in Australia.

In the U. S., concrete tile has been around since the early part of this century, but the industry languished until recently. In the early 60s, the Australians first brought their enthusiasm for concrete tile to California, along with machines for making high-quality extruded-concrete tiles. Since then, concrete tiles have become firmly established in the roofing market. Now more than 30 major plants and many smaller ones sell concrete tile throughout the U. S. and Canada.

The typical concrete tile (drawing facing page) measures 13 in. wide, 16½ in. long and about ⅝ in. thick. It weighs about 10 lb. Along both edges runs an interlocking channel, and across the front, underneath, are two or more transverse baffles that rest on the tile below and block blown rain and snow. At the back, projecting lugs allow the tile to be hung on roof battens. The surface texture is either orange-peel smooth or, for tiles that are meant

to simulate wood shakes, striated. (For a look at how these tiles are made, see the sidebar on the facing page.)

Color was added to concrete tiles originally to mimic the terra-cotta of clay tiles, the grey-browns of wood shakes, or the grey-greens of slate. Most people still prefer these earth-tones, yet tiles are also available in striking blues, greens and even white.

Concrete tiles are colored by one of two methods. The first is to add iron-oxide pigment to the batch mix, which produces a uniform color all the way through the tile. A less expensive method is to coat the tile with a slurry of cement and iron-oxide pigment. Use of this technique also allows the manufacturer to add highlights of a second color, creating a shaded or variegated effect.

Regardless of the coloring method, tiles are additionally sprayed with a clear acrylic sealer. The sealer not only helps the tiles cure properly but also makes sure that any efflorescence (the white powder—free lime—that surfaces as concrete ages) is driven out the underside of the tile rather than out the top where it would spoil the appearance. As a side-effect, the sealer gives the tile a slight gloss. However, the gloss wears off in a few seasons, and the color softens to its true matte finish. Before or after the tiles are installed, they can be painted with standard acrylic paint to change their color, much like painting a stucco house. Painted tiles require periodic rejuvenation.

Installing the tile roof—Because tile lasts so long, all other components of the roof should be designed to give equally long service. Tile manufacturers all specify these details in their brochures. Some list minimum standards and secretly hope their customers will use better materials if they can afford them. Other manufacturers, who have seen their own tiles outlive the flashings and felts, now specify only the higher-grade materials—copper flashings, heavy felts (coated base sheets of more than 30 lb., built-up membranes or single-ply roof membranes), and copper nails. For slopes that are less than 3-in-12, tile roofing should be considered decorative only. The real roof is installed underneath.

The procedure for installing a tile roof varies somewhat with the tile pattern, and each manufacturer covers the specifics in their installation instructions. Basically, the roof is first felted, and horizontal and vertical guide lines are chalked to indicate the courses. Layout is critical because any deviations stand out against the pronounced vertical pattern of most tile roofing. If the tile is designed with lugs to hang on battens, those battens are nailed on now (some manufacturers approve of hanging their tiles on spaced sheathing over heavy felt underlayment draped over the rafters). Then the tiles are loaded onto the roof so that they are evenly distributed and within easy reach.

The first course rests on either a raised fascia, a cant strip, special under-eave tiles or, in

Low-profile concrete tiles. **After the roof is felted, horizontal battens are affixed to the roof sheathing. Lugs at the top of the tiles bear against the edge of the battens to hold them in place. In seismic zones and windy areas, the tiles should also be nailed down.**

Making concrete tile

Back in the 1920s, commercial concrete tiles were made by hand. Workers would pour a "moderately stiff paste" of mortar into wooden or metal molds and then strike off the surface with a profiled screed. The green tiles cured in the molds for a week or more before they were separated and shipped. Today the process is automatic. To see the process firsthand, I visited the SpecTile (now Monier) tile plant in Salem, Oregon.

As I watched the blue Skandia tile extruder whir and thunk, then general manager Rick Olson explained to me what was going on. A conveyor belt fed concrete into the hopper of the extruder. The concrete was not the slurpy mix I have poured for walks and walls but a rich, dry mix (5% to 8% water; 20% to 25% cement) that felt more like castle-building beach sand. The dry mix makes a stronger tile that cures more quickly. However, such a dry mix cannot be worked by hand to the required density. That takes high pressure from the extruder.

The extruder plopped a measured amount of concrete mix onto a steel tile mold, called a "pallet," formed to produce the complex profile of interlocking ridges, baffles and lugs on the underside of a tile. The top of the tile was rolled roughly to shape. Then, a cam grabbed the bottom of the pallet and forced it under tremendous pressure beneath a forming die that shaped and troweled the top profile. As the continuous mat of pallets emerged from the extruder, a guillotine knife sliced between the pallets to separate them into individual tiles (photo right) while another arm punched holes for the nails. Their Swedish-designed, automatic tile-making system had been rated at one tile per second, but Rick told me they were pushing the equipment to 72 tiles per minute and that the extruder might be capable of cranking them out even a little faster.

Rick stopped the machine and pulled a newly formed tile off the tie line. "Here's one way to tell a quality tile," he said, pointing to the interlocking edge. The baffles and interlocking edges were crisp, knifelike edges.

The line of palleted tiles, resembling a nearly gridlocked freeway at quitting time, slid along tracks to the coating machine.

Photo: J. Azevedo

There, a pigmented cement batter drizzled onto a spinning horizontal brush, which flicked the slurry onto the tiles passing underneath. The dry-mix tiles sucked up the slurry, bonding the color to the surface. Another coating machine added a highlight color, or "flash," to the tiles. Finally, the tiles were sprayed with an acrylic sealer.

At this point the tiles had their final shape and color, but they still needed to be cured. We walked around back of the line to the insulated curing chamber, a space perhaps the size of an average house, warmed by the heat of curing concrete. Normally tiles spend 12 hours in this womb before being pushed out once more into the world.

Out of the curing chamber, the tiles rode the conveyor to the "depalletizer," where the pallets were separated from the tiles and returned to the extruder for another cycle. The cured tiles continued down the conveyor to where they were inspected for defects, stacked on recyclable wooden shipping pallets, shrink-wrapped, and moved out to the yard with a fork-lift. Olson, obviously proud of the company's quality control, told me that they must reject only one half of 1% in the plant, most of those for color problems.

As we walked around the yard, among white plastic cubes of bundled tiles, I asked Rick about business. "Well, these are all sold. I guess we could make more, but our employees like to spend nights and weekends with their families," he joked. And what about problems? "Our biggest problem is finding enough qualified installers who are willing to do a quality job." —J. A.

---

the case of mission-barrel tiles, a "birdstop" that also plugs the curved voids under the eave course (top photo, p. 63). Once the eave course is laid loose, checked for fit, and then nailed, successive courses are lapped over the one below as the roofer works diagonally up the roof.

Before the turn of the century, tiles on European roofs were fastened with two oak pins. Nowadays, nails will do just fine. The nails should be corrosion resistant at least; copper is better, especially in areas near the coast or affected by pollution. With standard concrete tiles on low-pitched roofs (less than 5-in-12) with battens, the field tiles need not be nailed. Their weight alone will keep them in place. The nailing schedule increases with pitch. In areas with high winds, all tiles are nailed and their butts secured with metal clips (available through your tile supplier).

With mission-barrel tiles, the fastening is a little more complicated (drawing next page). The trough tiles rest on the sheathing, so they can be nailed, but the cap tiles ride on the trough tiles, well above the sheathing. These tiles are generally fastened with copper wire that is either nailed back to the sheathing or tied to vertical copper strips running below

The anchor lug at the top of the tile rests on furring strips or sheathing, while the baffles at the bottom of the tile match and engage the lower course of tiles, preventing rain or snow penetration. The interlocking channels intercept rain blown from the side and direct it down the roof.

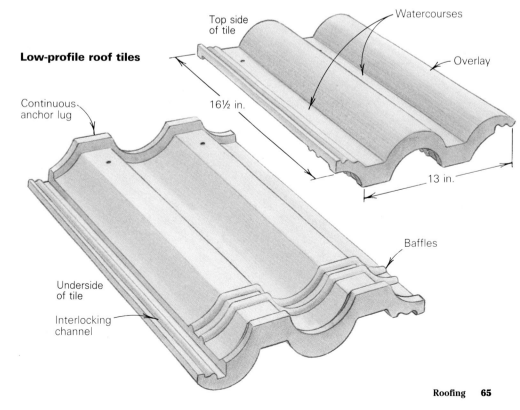

Low-profile roof tiles

Top side of tile

Watercourses

Overlay

Continuous anchor lug

16½ in.

13 in.

Underside of tile

Interlocking channel

Baffles

the tiles. Another method is to use tile nails, some of which look like gate hooks. Still another option is to nail vertical battens to the roof and nail the caps to these raised strips.

At the ridge or hip, concrete tiles are butted against a raised stringer. The joint is sealed with a flexible flashing, such as lead, and capped with ridge trim tiles affixed to the stringer. In place of flashing, some roofers prefer to point mortar under the edges of the ridge tiles to close off the watercourses. Care must be taken not to pack the joint fully because the mortar would then wick water up into the ridge. Clay tiles may be detailed the same way although some manufacturers offer special tile fittings to close the ridge and hip. With mission-barrel tiles, ridge and hip caps are usually mortared down. In heavy snow areas a tile roof, like any other kind, needs either extra felt or flashing that extends 2 ft. above the line of the exterior wall, or a ventilated roof deck (sometimes called a cold roof) to prevent damage from ice damming.

Once a tile roof is laid, it should need little or no maintenance. In fact, it is best to stay off of it. While tiles are strong enough to support careful foot traffic, someone clomping around on the roof risks breaking a tile. If you must walk on a tile roof, pick your way carefully across the reinforced or supported section of each tile, toward the nose (over the lugs) of concrete tiles and in the troughs (or "pans") of clay tiles. Mortared hips and ridges also

make good paths. Consider laying down a sheet of plywood to distribute your weight, or fill burlap sacks with sawdust and walk on these "pillows."

The question of weight—Try this experiment. Go up to one of your friends and say, "I'm thinking about putting on a tile roof." Chances are your friend will reply, "But aren't they heavy?"

The answer to that question is, of course, "Certainly." A typical tile roof weighs about 900 lb. per square—at least three times as much as wood or composition shingles. In the past, tile's reputation for being heavy has discouraged its use, but a heavy roof is not necessarily a bad roof. In the case of new houses, most are designed to carry three composition roofs: the original plus two re-roofs at intervals of 15 years. A tile roof weighs about the same. For new construction, then, a tile roof will not require any special framing. If you like tile, don't let its weight deter you.

When considering re-roofing an old house with tile, an owner should consult a building inspector or engineer to see if the structure was designed to handle the weight. While most houses can support a tile roof, an occasional old house that was built to minimum standards may have to be reinforced before tiling. Usually, this reinforcement takes the form of a braced purlin at mid-span, and the structural work adds a few hundred dollars to the

roofing expense. An alternative would be to choose one of the lightweight-concrete or hybrid fiber-cement tiles on the market. At 40% to 60% of the weight of standard concrete or clay tile, these materials generally need no additional bracing. But watch the costs. Bracing can be cheaper than the premium charged for these lightweight materials.

During an earthquake, the rolling and shaking of a house has been known to flip unsecured tiles off the roof. Also, on tall houses especially, the weight of any heavy roofing sets up extra stress in the frame as the house whips back and forth. For those of you building in earthquake country, consider taking the precaution of securing at least every other row of tile with nails. And make sure that the house has enough shear strength to resist wracking. Check with your local building department or consult with a structural engineer if you are in doubt about the ability of your house to bear the weight of a tile roof during an earthquake.

I was curious about the effects of the 1989 earthquake in Santa Cruz, California, on tile roofs, so I called the building department there to find out how tile roofs fared during the shake. I was told that, in general, tile roofs held up well. Any tiles that were shaken loose were unsecured and were typically on roofs steeper than 5-in-12. Tall houses with tile roofs showed no structural damage, reflecting the soundness of the tough engineering requirements that have been in force there for some time.

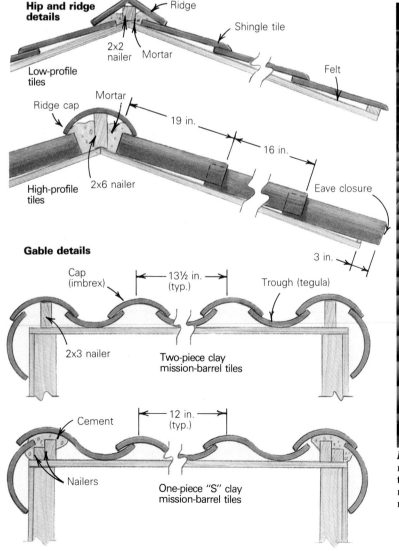

Hip and ridge details

Ridge

Shingle tile

2x2 nailer

Mortar

Low-profile tiles

Felt

Mortar

Ridge cap

19 in.

16 in.

High-profile tiles

2x6 nailer

Eave closure

3 in.

Gable details

Cap (imbrex)

13½ in. (typ.)

Trough (tegula)

2x3 nailer

Two-piece clay mission-barrel tiles

Cement

12 in. (typ.)

Nailers

One-piece "S" clay mission-barrel tiles

High-profile clay tiles. **Tapered two-piece mission-barrel tiles have nail holes at opposite ends, depending on whether they are cap or trough tiles. The cap tiles above are secured with long nails. It's best not to walk on a tile roof—all the cap tiles behind the broken cap in row five will have to be pulled before it can be replaced.**

What will it cost?—Not surprisingly, the cost of a tile roof will depend on the type of tile, the distance from the manufacturer to your site, the complexity of the roof, the quality of installation you specify, and who installs it. Like other premium roofing products, such as slate and metal, tile costs more initially than an ordinary composition roof but can have a lower life-cycle cost because it lasts much longer. Also, the appreciation in property value and lower maintenance must be figured into the calculation.

Concrete tile generally is the least expensive type of tile roofing. A basic concrete-tile roof installed by a knowledgeable roofer in an area where tile is common, such as the Sunbelt states, will cost about $125 to $150 per square (100 sq. ft.), with about half of that as materials and the rest labor. In an area where tile is still a new idea or with a roofer just getting started in tile, the cost will be higher. A complex roofline will cost more for both labor and materials. Premium flashing and fasteners will also add to the expense.

In the South and Southwest, the home of clay tile, a new roof of unglazed, one-piece mission tile will cost the same or a bit more than concrete. Two-piece mission-barrel tiles run a little more because there are more tiles to install. For most other areas of the country, where clay-tile plants are scarce, shipping adds considerably to the cost.

Glazed clay tiles stand on the top rung of the cost ladder. The tiles alone start at just under $170 per square—about the same as treated shakes—with shipping, fittings, labor and profit on top of that. The most expensive glazed tiles that I've heard about top $1,000 per square. Most people who are contemplating a tile roof cannot justify the additional cost of glazed tile versus unglazed. There are times, however, when least cost is not the object, and intangibles enter into the equation. The rich color and texture of some of the glazes on classic profiles may seduce an otherwise analytical voice that speaks of budgets and mortgages.

Finding the right tile—Your first step in selecting a roofing tile is to reconcile your taste with your budget. Once you have decided on a general category of tile, you will want to find a reputable manufacturer. Concrete tiles are now manufactured throughout the Sunbelt states, and plants are beginning to spread across North America and Hawaii. Clay tiles come from plants in California, Florida, Ohio, and New York. Glazed tiles are made in California and Ohio, and are imported from the Orient to West Coast ports. To find a nearby manufacturer, check with your local roofing-supply house. If you cannot find the tile you want, call or write the National Tile Roofing Manufacturers Association (NTRMA, 3127 Los Feliz Boulevard, Los Angeles, Calif. 90039; 800-248-8453). Tell them what you are looking for, and they will direct you to a manufacturer or help you find the tile that is right for your project.

Before you buy, check into the reputation of the manufacturer. If you choose a tile from a member of the NTRMA, the tiles will meet or exceed ICBO (International Conference of Building Officials) standards and have a 50-year warranty. If you choose a tile made by someone else, ask about their warranty and for a copy of their ICBO report. Whatever you do, walk away from a supplier who cannot or will not provide an ICBO report that matches the identifying marks on the tile.

As important as finding the right tiles is finding the right person to install them. The place to begin is with the manufacturer. Most manufacturers keep a list of roofers who have been trained to install roofing tiles. When you talk with the roofer, start out with the standard questions about experience, licenses, insurance, and references. Then together go over a copy of the manufacturer's installation guidelines. If the roofer suggests varying some detail, check that it will not void the manufacturer's warranty. Ask the roofer about upgrading the flashings, fasteners, and felts. It seems false economy to scrimp on the details after committing to the investment in a roof that will outlive the next generation. Here may be the one insoluble disadvantage of a tile roof: it forces us to think beyond our own mortality.□

J. Azevedo is a free-lance technical writer, working in Santa Clara, California.

Glazed clay tiles. Tiles made of clay can be finished with baked-on glazes in both custom or stock colors. This Japanese-style roof in glossy white is essentially the same as the interlocking S-type mission tile, but its trim pieces, such as the cylindrical hip starters and the belled ends of its hip and ridge tiles, give it an exotic flair.

While most tile roofs cap a house in earth-toned silence, the rich colors made possible by using vitreous glazes on ceramic tiles can make the impact of the roof the dominant element in the bearing of a house.

Putting on a Concrete-Tile Roof

It takes nimble feet, a strong back and attention to flashing details to do it right

by J. Azevedo

Most people with even a little construction experience would not hesitate to install an asphalt-shingle roof, yet they quake at the idea of putting on concrete tile. Maybe it's the mystery of flashing, where the undulating tile profile meets the straight line of ridge, or worse yet, the angled line of hip or valley. Maybe it's the uncertainty of laying a material that won't bend, not even a little, and is too thick to slip under a skylight flashing. Or, maybe it's simply the idea of moving all that weight around. After all, an average roof might need 16 tons of tiles, and that number carries with it the image of hauling coal, growing older daily and falling deeper in debt.

After presenting the case for choosing a tile roof in another article (see pp. 62-67), I started to wonder whether there was in fact a bona fide basis for the amateur's reluctance to roof with concrete tile. To find out whether it's a job best left to the professional, I went up on the roof with one: Marlen DeJong, principal in the modestly named A-1 Tile Company of Everson, Washington. After spending a week watching him install roofs on three houses over the course of a summer, I have concluded that tile roofing, though physically demanding, should not intimidate a conscientious builder. The interlocking tile is virtually foolproof. The only tricky spots—and these do require careful attention to detail—are the "joints" in a roof: the ridge, rake, valleys, hips and roof penetrations.

Readying the roof—A tile roof requires a little more preparation than a shingle roof. Special attention must be paid to setting fascia, applying felt underlayment and nailing on horizontal battens. DeJong asks the carpenters to run the eave fascia ⅝ in. past the rake and to set it 1½ in. above the plywood roof sheathing (top photo, facing page). Like a shingle starter course, the raised fascia lifts the first course to the same pitch as the other courses.

The felt underlayment forms a second line of defense should any water somehow sneak past the tile. DeJong uses 30-lb. roofing felt, 36 in. wide, lapped 3 in. at each course. At the eave, where the raised fascia would dam the water, he installs an "anti-ponding strip," which is a preformed sheet-metal flashing (available from most roofing-tile manufacturers) that takes the felt (and any runoff) smoothly over the raised fascia. In areas that get snow, Dejong doubles the felts along the eaves and seals their edges with roofing tar to seal out any water that might get beneath the tiles as a result of ice damming.

At the ridge, DeJong normally takes the felt over the peak and down the other side a foot or so. If there are ridge vents, however, he holds both the sheathing and the felt back 2 inches from the peak. At valleys, he runs the felt from one slope across the valley and a few feet up the other side and cuts the opposite felt to lie right down the center of the valley (variations include interlacing the felts, and even adding another strip vertically down the valley). DeJong lays standard valley flashing on top of the felt (middle photo, facing page). The flashing is 24 in. wide and shaped like a "W" in profile, with the long edges turned up and inward to contain runoff. The flashings are secured with 8d nails bent over their long edges. Where valley flashings overlap, the bottom piece can be nailed near its top edge. The next piece overlaps it by 6 in., and the joint is sealed with roofing cement. Where the valley flashing meets the anti-ponding strip, he crimps the valley to match the change in pitch.

Felts and flashings are the weak spots in a roof that is supposed to last better than 50 years. DeJong offers his customers upgrades—copper or lead flashings instead of painted galvanized steel; double felts or 45-lb. felt instead of 30 lb.—but almost no one goes for it, an attitude that seems shortsighted.

Battens—With the roof felted, DeJong starts to lay out the battens—the horizontal strapping (typically 1x4s or 1x6s) that will support the tile. The battens define the course spacing, so the layout must be precise (bottom photo, facing page). First, he staples ¼-in. by 2-in. lath over each rafter. The lath raises the battens, allowing any moisture to drain down to the gutters. Then he sets the first batten parallel to the eave, with its uphill edge 14½ in. back from the front edge of the eave fascia. He goes then to the ridge and sets a batten 1¾ in. down from the peak. Next comes a calculation to determine the course spacing. He divides the distance from the ridge batten to the eave batten by 13½ in. (the ideal course). If the result is not a whole number (and it seldom is), he takes the next highest number of courses and divides the distance by that number to get the course spacing. If you don't like math, don't worry. Tile manufacturers provide tables that give course spacing for any roof. Once DeJong has the spacing, he marks out the courses on the felt, snaps chalklines, and nails down the remaining battens, staggering the joints between them.

The only complication in laying the battens is when two roof slopes of unequal length meet as one at the ridge, such as at staggered eaves. In that case, the course spacing for one roof must be used on the other side as well. That usually results in a short course at the eave.

Loading the roof—With a roof that averages 16 tons, three things are certain: you don't want to carry the tiles up a ladder; you don't want to stack them all in one place; and you don't want to move a tile very far once you put it down. Fortunately, roofers have worked out a system for avoiding the three "don'ts."

Tiles arrive on pallets, and suppliers will either deliver them to the roof or arrange delivery at extra cost. From there it is up to the roofer to unload and distribute the tiles.

DeJong guides a pallet, swinging from a boom truck, to a spot near the ridge and has it lowered until the front just rests on the roof. He dons a pair of heavy gloves and unloads a bundle of six tiles near the peak. He continues this way along the ridge, spacing the bundles 2 ft. o. c. Then he moves down three battens and lays another row of bundles. He works his way down the roof; the last row of bundles rests on the third batten up from the eaves. Any extra tiles go back up near the ridge or near hips.

Before stacking the roof, DeJong looks at the batch dates on the pallet wrappers. Each batch may vary slightly in shade, so he keeps all tile from the same batch on one section of roof. Other roofers like to distribute tiles to mix the variations into the roof.

Laying the field—The large interlocking field tiles go down without much trouble and cover a roof quickly. DeJong starts from his left as he lays the first eave tile (bottom photo, facing page), hooking the lug on the bottom of the tile over the first batten and holding it back from the rake about 1 in. to 2 in. The next tile interlocks with the first along their common edge (called the "waterlock"). As he continues along the eave, DeJong checks the waterlock on each tile to make sure it's clean.

Sometimes a bit of the slurry color coating on the tile has clogged the waterlock, and he must scrape this out or the tiles will not fit neatly. He lays them loosely, with a gap of 1/32 in. to 1/16 in. to allow for structural movement. When he gets to the opposite rake, he hopes to end the course with either a whole tile or a half tile. If this doesn't happen, he can usually adjust the course, squeezing the tiles or pulling them apart slightly to fit. The goal is to have one of the three tile "barrels" (the ribs or humps of a tile) at each end of the course to direct water away from the rake, toward the watercourses on the field tiles. When he is satisfied with the fit, he drops an 8d galvanized nail through the hole in each tile and taps it snug. How snug? "If you break the tile, you know it's too tight." Yet, you don't want to leave it sticking up. If you do, it will pop the corner off the next tile up the first time you walk on it.

Once the first course is nailed, DeJong measures in from the rake to the second joint and makes a mark. He marks other joints about every 10 feet along the course, and then snaps parallel lines all the way to the ridge. These marks will serve as guides to keep the courses aligned vertically up the roof.

DeJong lays the second course much the same as the first, only he starts with a half tile instead of a whole one (to learn how to cut and break tiles, see the sidebar on p. 73). He continues this way, course by course up the roof, staggering the joints and lining them up to the chalk marks, thumping each tile with the rubber handle of his hammer and listening for the telltale rattle of a tile that is not lying flat against its neighbors or the batten. As he falls into the rhythm of laying field tile, he seems to deal out these 10-lb. tiles with all the ease of a casino blackjack dealer.

On roofs with less than a 12-in-12 pitch, he nails only every other course. Above that, or if called for by the manufacturer's specifications or local codes, he nails every tile. On really steep pitches or in high-wind areas, he also fastens down the nose of each tile with a copper hurricane clip (available through the tile supplier).

Hips and valleys—Hips and valleys are basically mirror images and the tiling is much the same. It's a matter of getting the angles right and holding the cut pieces in place.

To tile a hip, DeJong starts by toenailing a 2x4 nailer on edge along the hip, holding the lower end back about 4 in. to 5 in. from the eave. Then he lays field tile as close as he can to the nailer without cutting any tiles. Near the nailer, he can use broken tiles that he has been saving. After DeJong runs a circular saw along a chalkline snapped parallel to the hip (bottom left photo, next page), he blows off the dust and tosses the scrap. The cut pieces should fit tightly against the nailer (top photo, next page) once he inserts two standard field tiles into each course.

DeJong fastens the cut pieces two different ways, depending on their size. For large pieces,

At the eave, a raised fascia supports a sheet-metal "anti-ponding strip." The strip acts as a starter course to raise the level of the first course of tile into plane with the rest of the roof.

Valley flashings are 2 ft. wide, with an angled lip along both edges that directs runoff back toward the flow line. DeJong secures the flashings to the roof with 8d galvanized nails driven outside the flashing edges; he then bends them over the lip.

Before the battens go down, DeJong staples lath to the deck over each rafter. This allows any moisture that may sneak under the tiles to escape easily. Here DeJong begins the first course of tile, hanging tile lugs on the first batten.

which still have part of the lug, he hooks the tile over the batten, drills a hole near the top, and nails it in place. For small pieces, he drives a nail partway into the deck near the nailer as a platform to support the edge of the tile. He hooks the bottom edge with another nail driven sideways into the nailer.

If a small angled tile finishes off the eave course where flashing prevents him from driving nails, he cements the piece to the next whole tile with sticky black roof tile cement (Fields Corp., 2240 Taylor Way, Tacoma, Wash. 98421; 206-627-4098) in the top 2 in. of the waterlock. Once the tiles are set, he then tapes the two tiles together to hold them until the cement sets up. DeJong flashes the joint by molding 4-in. flexible flashing tape (Aluma-Grip-701, Hardcast Inc., Box 1239, Wylie, Texas 75098; 214-442-6545) down from the top of the nailer and onto the tile.

Finally, he caps the hip with special trim tiles (small photo, below left). He test-fits a hip tile, like putting a saddle on the nailer, and checks that the bottom edges just rest on the barrels of the adjacent field tiles. If the nailer is too low, he shims it with lath. Then he starts the hip with a bullnose end cap nailed to the nailer. He spreads tile cement over the nail, drops a standard hip tile down on the nailer, slides it down into the cement until the bottom lug catches on the tile below and nails it in place. He continues up the hip to the apex, where he either miters the hips to meet the ridge (a ridge tile will later slip under this mitered joint) or caps it with a special apex tile. In either case, he flashes under the apex with flexible flashing or lead.

Some roofers prefer to bed the hip tiles in mortar. If you choose to do this, be sure not to pack the joint. Just tuck mortar under the edges. Otherwise, the mortar will wick water up into the hip.

Like hips, you can cut and fit valley tiles one at a time or cut the whole row at once. But when it comes to securing cut tiles to the roof, DeJong alters his technique. For large pieces, he strings a length of 14-ga. galvanized wire through the nail hole and ties the wire back to a nail driven just outside the valley (top photo, facing page). In this way he avoids punching holes in the valley flashing. If any of the cut tiles tips or rocks, he sticks a chunk of broken tile underneath to support it (some roofers tuck mortar under the cut edge).

Rakes and ridges—Rakes are trimmed with special tiles that are nailed to the rake fascia. Rake tiles resemble hip tiles but are formed to a tighter "V." They are interchangeable for right and left rakes.

Balancing precariously on the eave corner, DeJong sets the first rake tile on the end of the eave course and slides it up against the nose of the next field tile up. He holds the side of the rake tile plumb and carefully drives two nails through the prepunched holes. Working his way up the rake, he butts a rake tile against each course of field tiles until he gets within

Hip trim. It's most efficient to gang-cut angled tiles for hip or valley intersections. Note in the foreground (left photo) the position of the nail holding the chalkline. The distance from it to the corner of the tile to its right is the same as the distance from the tile nearest the hip to the edge of the nailer. Trimmed hip tiles tuck against the 2x4 nailer (top photo). Short pieces are supported at their lower edges with a nail driven horizontally into the 2x4. Asphalt roofing cement, along with flashing tape, can also be used to glue smaller tiles to their neighbors. Hips are finished with trim tiles affixed to the 2x4 nailer with 8d galvanized nails topped with a dab of asphalt cement (photo above). The shim under the second trim tile ensures a sturdy base for the tile.

No holes in the valleys. At valleys, tiles are secured without penetrating the flashing. On the near side, a tile is wired to a batten. On the far side, DeJong tapes a piece of tile to its neighbor. Asphalt cement between the short piece and the overlapping tile will further secure it.

Rake tiles that are V-shaped in section trim the gable ends of roofs. At the top, they are mitered and sealed with asphalt. Trim tiles similar to the hip tiles finish off the ridge.

DeJong's preferred method to seal chimneys and sidewalls against the weather is with lead step flashings shaped to fit the irregular contours of the tile.

Another way to tile up a sidewall, chimney or skylight is with a sidewall flashing. Tile edges overlap the raised lip of the sidewall flashing, spilling runoff into the flashing channel.

A vent is sealed by folding the site-cut tabs of a lead jack over the lip of the vent.

tangle of 2½-lb. lead sheeting and bends it in half lengthwise. He spreads tile cement on the bottom half and presses the flashing into the corner, with the bottom edge even with the headwall flashing. The lead can be molded to conform to the shape of the tile below.

DeJong prefers the step-flashing system for sealing along chimneys or sidewalls (bottom center photo, previous page). He lays the next tile course right up to the side of the chimney and sets another bent, gooped, 12-in. by 16-in. piece of lead step flashing atop that tile, with the nose of the flashing even with the nose of the tile.

Once he works flashing up to the top of a skylight or a chimney, DeJong fits in a piece of rigid "back flashing"—a sheet-steel "V" that extends up the back curb of the skylight and 16 in. up the roof. At the sides, the back flashing extends 6 in. past the edge. DeJong blocks under the flashing with battens and tile pieces to raise it up just high enough that it sets on top of the step flashing. To make the joint, he cuts down the crease of the last step flashing almost to the corner of the curb and folds the vertical flap around the corner and against the back curb. He also cuts down the top flap of the back flashing to the corner and bends it around the corner and against the side. When the fit is right, he pulls the back flashing, goops all the overlapping flaps and corners with roof tile cement and presses the back flashing back down into place. The next course of tiles runs across the top, sitting on the back flashing.

If the skylight or chimney is wider than 4 ft., DeJong hires a carpenter to frame a cricket. Then he tiles around it as he would a valley.

The other way to flash the side is with a single piece of sidewall flashing (sometimes called J-flashing). This is a sheet-metal pan turned at a right angle to go up the wall. It has a raised lip along the outer edge of the pan like valley flashing. The flashing rides on battens, and the edges of tiles drop into this pan, which channels water down the side and spills it onto the course below the chimney or skylight.

At least one skylight company (Velux-America Inc., P. O. Box 3268, Greenwood, S. C. 29648; 800-888-3589) offers its own tile-compatible flashing kit using the sidewall flashing approach (bottom right photo, previous page). If you buy these skylights, you should probably consider getting the flashing as well. It's not hard to install, despite the instructions.

The flashing at the base of the Velux skylight shown here is made of ribbed lead sheeting. Before one can be installed, the protruding barrels of the tiles need to be lopped off so the sheet will lie flat. Also, the noses of the side tiles are notched to clear the lip of the sidewall flashing, and the tiles are tied back with wire to avoid puncturing it.

Tiling around a vent pipe—Flashing around a plumbing vent is quite a bit easier than sealing larger openings. The secret is the malleable lead jack (photo above).

DeJong tries to coordinate his work with the

one tile of the ridge. Ridge and rake will meet at the apex, so he first must prepare the ridge.

Along the ridge he has already toenailed a vertical 2x3 nailer, stopping it back 6 in. from the gable end. The top course of field tile nearly butts this nailer. Now he molds flexible flashing tape down the nailer and onto the tiles, as at the hips, to seal the ridge joint. At this point he can cut the apex of rake and ridge.

DeJong miters the opposing rake tiles at the peak so that they meet with a neat, tight joint. He drills another nail hole in each trim piece, sets nails in the rake as spacers to hold the pieces plumb and square to the rake, and nails the mitered trim with two nails each.

To cap the ridge, DeJong butts a ridge tile square against the mitered rakes (bottom left photo, previous page) and nails it to the 2x3 nailer, as at the hips. Another way to do this detail is to notch the bottom of the ridge tile so that it overlaps the mitered rake joint. He goops the mitered seam and stuffs more tile cement into the arch of the ridge tile where it meets the rake. At first the shiny black cement

looks unsightly and out of place, but it will eventually fade and the color then will blend with the tile.

DeJong completes the ridge by lapping trim tile, nailing and cementing the joints as he did on the hips. When he gets close to the opposite end, he checks the distance remaining and adjusts the spacing to get the last ridge tile to meet the apex of the rake. Or, he simply cuts the last tile to fit.

Chimneys and skylights—DeJong tiles right up to the base of a chimney or a skylight as if it were a ridge. If it works out that this last course is short, he cuts the tiles and uses the bottom portions, drilling them for nails and shimming them as necessary to maintain the roof pitch. Then he seals the joint with Aluma-Grip tape, and caps that with rigid counter-flashing (sometimes called "headwall flashing"), which extends down 6 in. over the barrels of the tiles below. Incidentally, he uses this same detailing where a roof meets a headwall.

To flash a bottom corner of a boxed-in chimney, DeJong cuts a 12-in. by 16-in. rec-

plumber. Ideally, he would like the vents to come up between the battens, in the center of a tile. This location produces the neatest, most water-tight job with the least fuss. The vents should project about 15 in. above the deck.

As he tiles up to the vent, DeJong measures, marks, and cuts the tile to fit around it, and then sets the tile in place. He slides a lead roof jack over the vent and onto the top of the tile, square to the roof. With the rubber handle of his hammer, he dresses the jack to conform to the curves of the tile, and he trims the back of the flashing even with the back of the tile. He slides the jack up, goops roof-tile cement around the hole in the tile, and slides the jack back down, bedding it in the cement. Then he folds inward site-cut tabs in the top of the jack to seal the vent.

Safety—To help you keep your footing, DeJong recommends you wear rubber-soled shoes (athletic sneakers or deck shoes). Concrete dust is slick; blow off the roof after you cut tiles. And, stay off a wet roof.

On steep pitches, the situation gets more precarious, and safety becomes more critical. For modestly steep pitches, say 8-in-12 or 12-in-12, stay off the tiled portion, walk on the battens and make "ladders" to get across tiled slopes. Steeper than that, plan to hang on the batten with one hand, and consider a scaffold or safety net at the eave to give you a second chance. Or, maybe this is the time to call a professional. The National Tile Roofing Manufacturers Association (NTRMA, 3127 Los Feliz Boulevard, Los Angeles, Calif. 90039; 800-248-8453) makes available a list of tile manufacturers, who can in turn recommend competent installers. □

J. Azevedo overcame his acrophobia to research this article. Photos by the author.

Cutting and breaking concrete tile

Concrete tiles are only sand and cement; you can cut them with a carborundum wheel mounted on your circular saw. Still, abrasive masonry wheels are exasperatingly slow, and you will wear out several on a tile roof. If your roof is cut up by a lot of hips, valleys, and skylights, DeJong recommends that you consider buying a diamond blade. He cuts tile with a 12-in. diamond blade mounted on a portable electric cut-off saw (for more on these tools, see *FHB* #62, pp. 80-84). The diamond chews through tile about as fast as a circular-saw blade cross-cuts 2x fir.

Breaking tiles is easy, but getting them to break where you want takes a little more finesse. DeJong generally breaks, rather than cuts, half tiles for starting courses at the rake. First he cuts or chips off the lugs and reinforcing ribs down the center of the underside of the tile. Then he cradles the tile, face side up, in his arms like a baby and raps the center with a hammer. If all goes well, and it generally does, the tile splits in half down the center (photo below right).

To drill a small hole in a tile, DeJong uses a standard masonry bit. For larger holes, as for vent pipes, he scores a square with his diamond blade and knocks out the opening with his hammer. —*J. A.*

After laying out a chalkline parallel to the hip, DeJong uses a 12-in. dia. blade to make clean, fast cuts in the concrete tile.

To divide a tile without a saw, remove the lugs from the center of the tile's underbelly, flip it over and give it a rap in the middle with a hammer.

'If I ever become a rich man,
Or if I ever grow to be old,
I will build a house with deep thatch
To shelter me from the cold...'
—Hilaire Belloc

Thatching

This ancient craft is still practiced in England, where they know their yealms from their nitches

by Alasdair G. B. Wallace

Thatching is one of Britain's oldest building crafts. Long before the Norman Conquest, the Saxons referred to any kind of roofing as *thaec;* the act of applying it was *theccan.* Historians agree that the Neolithic stone circles were roofed with locally available materials— turf, heather, bushwood and reeds supported by a pole framework. The craft of thatching and its tools have changed little since the Middle Ages.

The advent of the railway in the early 19th century, with the concomitant growth of urban centers, made slate, tile and iron roofing materials readily available. Thatching declined in popularity for the next 100 years, although as recently as the 1950s, the village thatcher's work was commonly found in ricks, stacks, outhouses and cottages. Today, though, the modern combine harvester effectively destroys the material on which the craft of thatching is based. There is also an increasing reliance on high-yield wheat strains, which

For hundreds of years, bundles of combed wheat like this have been transformed into efficient roofs for small buildings by highly skilled English craftsmen.

produce shorter, weaker straw, and the traditional machinery for processing the product is becoming scarce. As a result, thatch is rapidly becoming an exclusive and expensive status symbol—an ironic fate for a material that was used for centuries because it was commonly available. Generally speaking, the initial cost of thatching a building in England today is between three and five times greater than that of roofing it with a conventional material like slate or tile.

In the minds of most urbanized Westerners—and many country people as well—a thatched building imparts a sense of timelessness, of the rural tradition, and of serenity. Thatching also provides fine insulation against heat, cold and noise. No two thatched buildings are alike; the craftsman's material is almost infinitely malleable, and responsive to his every whim. As you can see in the photos on the facing page, attic windows, dormers, gables, ridges and gazebos invite artistic li-

---

Thatching nomenclature

Yealm: A bundle of long straw, about 5 in. thick and 14 in. to 18 in. wide. The basic long-straw thatching unit.

Bottle: A double-thickness yealm, tied with twine or tarred cord at the small end. Bottles form the first-laid courses of wheat reed on eaves and barges.

Nitch: The standard unit of combed wheat reed, the butt ends of which lie in one direction. It is usually tied with binder twine. The minimum permissible reed length is 27 in.; 36 in. is ideal. One nitch weighs about 28 lb.

Wadd: A double handful of combed wheat, tied together near the top. Wadds are used like long-straw bottles at the eaves and barges.

Bunch: A unit of water reed, which is graded according to length. Long reed is 6 ft. or more

and used at the eaves; short reed is 4 ft. to 6 ft. long, for use at the barges; coarse reed is 3 ft. to 4 ft. long, and used for general-purpose work. Water-reed bunches are about 24 in. in circumference at a point 12 in. from the butt end.

Butting: The act of dropping the butt end of wheat or water reed on a hard, flat surface—usually a butting-board—in order to produce a dressed nitch or bunch.

Dolly: A roll of wheat or water reed 4 in. to 8 in. in dia., secured to the ridge before capping.

Sedge: An evergreen marsh plant used to cap water-reed roofs. It is often scalloped or trimmed with the small knife into an ornamental pattern.

Course: A single horizontal layer of straw or reed as it is applied to

the roof. The brow course is the first course laid atop the wadds or bottles.

Barge: Variously referred to as verge or gable, it is usually called the rake or gable-end overhang in the United States. In thatching, barge refers to the finished edge of the thatch that overhangs the barge-board.

Tilting board: Sometimes referred to as the fillet or arris. A board at the eaves and barge, which forces the butt ends of the reed upward, exposing them to the weather and also providing a desirable tension in the thatching material.

Sways: Hazel rods, sometimes split, 4 ft. to 10 ft. long, used to secure each course. Held by tarred cord to the rafters, or more commonly by iron hooks, sways are always covered by subsequent courses.

Liggers: Also called **rods,** they are made of split hazel or willow depending upon locale and supply. Four to five feet in length, they are used to secure the sedge at the ridge and long straw at eaves, barges and ridge.

Cross-rods: Ornamental split hazel rods, 12 in. long, that are used between liggers at the ridge.

Spars: Split hazel or willow rods, 30 in. long, that are pointed at both ends and bent and twisted at the center. In effect they are wooden staples, the ends of which are under tension when they are inserted into the thatch. They are used to secure liggers and cross-rods.

Netting: Applied to wheat-reed roofs as protection against birds and vermin. The most common types are ¾-in. galvanized or polyethylene.
—A. G. B. W.

Most thatching being done today is actually re-thatching. Here, new combed wheat is being applied over old. In the decayed older thatch, the sways—hazel rods hooked to the rafters to hold the courses of thatch in place—have been exposed. The thatcher works along the roof, completing one section at a time all the way to the ridge. Here, a wadd is in place at the eave of the last section.

The tilting board

The tilting board forces the eave course upward so its more durable ends are exposed. Hazel sways force the thatch into tension and hold it in place. There are several ways to install tilting boards, depending on the eave treatment that is appropriate for the house.

Full fascia

Sway
Rafter
Battens (about 9 in. o.c.)
Tilting board
12 in.

Soffit

Sway
Tilting board
15 in.

Exposed rafter tails

Sway
Tilting board
15 in.

cense. There are no eavestroughs (gutters) or downspouts to detract from the roofline or to clog with autumn leaves.

Thatching materials—Today, two types of material are commonly used for thatching: wheat straw and water reed. Wheat straw produces two varieties of thatching material: threshed long straw and combed wheat reed.

Threshed long straw is available only when the wheat crop is harvested by a binder, an antiquated machine (increasingly rare), that ties wheat into sheaves. Several of these sheaves are leaned together to form a stook. Eventually, the wheat is threshed to separate the grain from its stalk. Threshing crushes and splits the stalks, resulting in a haphazard pile of straw, with butts and tips intermingled. The butt end of the straw is the stronger and more weather-resistant part of the plant, so this process effectively halves the life expectancy of roofs thatched with long straw. This haphazard pile must be sorted into manageable yealms, the traditional term for a bundle of long straw about 5 in. thick and 14 in. to 18 in. wide. (For definitions of thatching terms, see the bottom of the previous page.) Long straw also requires constant moistening during application to render it flexible.

From a distance, a roof thatched with long straw has a rounded, poured-over-with-cream appearance. The eaves and barges (the British term for gables) are characteristically decorated with liggers and cross-rods, the thin hazel or willow wands that are fastened on top of the thatch to hold it down. The raised portion of thatch on either side of the ridge is also often elaborately decorated with an intricate pattern of these rods, along with scallops and geometric shapes in the thatch itself. This pattern is often the characteristic signature of the craftsman.

Combed wheat reed is produced when the grain is removed from the straw by combing rather than by threshing. In combing, the straw remains uncrushed and oriented with its butt ends all at one end of the nitch—a

bundle of combed wheat that weighs about 28 lb. The ideal length of the straw is 36 in.; 30 nitches will cover one square (100 sq. ft.). At one time, combing was done by hand, but today the reed comber, a specialized device attached to the top of the threshing machine, performs the task.

A roof thatched with combed wheat reed looks very different from one thatched with long straw. The ridge decoration is similar, but barges and eaves don't need to be secured with liggers and spars. The combed wheat roof has a more sharply angled, tailored look.

Water reed (Phragmites communis trin) occurs naturally in reed beds in England's lowland and coastal marshes. The strongest and most weather-resistant part of the reed is the stem. Since only the butt ends of the reed are exposed to the weather, this plant is an ideal thatching material. Reed bunches are graded coarse (3 ft. to 4 ft. long), short (4 ft. to 6 ft.) and long (over 6 ft.).

Reed is traditionally harvested in December or January, after the leaf has been killed by frost. The best reed is produced by regularly harvested beds. The marshlands of East Anglia and Norfolk have long been renowned for their reed, which in some areas is referred to as Norfolk reed. Today, though, three-fourths of the water reed used in the British thatching trade is imported from Europe.

Roofs thatched with water reed look like those thatched with combed wheat. Ridges are frequently decorated in a similar fashion, but eaves and barges are left unclipped. As with combed-wheat roofs, the thatch should be at least 12 in. thick; long-straw roofs should be at least 15 in. thick.

Durability—The longevity of a thatched roof depends on several factors: which material is used; the quality of the workmanship; the exposure, orientation and pitch of the roof; and how well it is maintained. But generally long straw will last from 10 to 20 years, combed wheat reed will last from 20 to 40 years, and water reed will last from 50 to 70 years.

Re-thatching (photo above) accounts for 80% of the trade's business these days in Great Britain. There is considerable variation in the amount of old material that the thatcher must remove before he gets down to thatch that is sound enough to act as the base for the new. A total thickness of 3 ft. is about the limit, both because of the material's weight and because it would be difficult to fasten sways to the rafters through a greater thickness. More on this part of the process later.

Water reed is much more brittle than either long straw or combed wheat. Consequently, new water reed will not mesh with old, and

Illustrations: Frances Ashforth, based on drawings in *The Thatcher's Craft*

roofs thatched with it have to be completely stripped before they can be re-thatched.

During the summer of 1983, I visited England and talked with several owners of thatched homes. Only once did the cost of maintaining the thatch become a topic of conversation. In this instance, an elderly widow mentioned that she had last had her home re-thatched in 1930 with Norfolk reed. She did not remember the cost, but hastened to point out that the roof is as sound today, 50 years later, as it was the day the thatching was completed. What other type of roofing, she wondered, would have lasted so well?

The roof structure—Most buildings being thatched today in England are hundreds of years old. When they were built, available materials were used as dictated by common sense and a few basic principles. Diversity flourished; ingenuity abounded. None of the roofs I inspected on my trip to England last year employed any external sheathing. All were of a standard rafter construction like that used in the United States and Canada, but the rafters themselves were a different story: pole, dressed, roughsawn, pitsawn, adzed, you name it. In place of sheathing, 1x2s or something similar had been nailed across the rafters. The thatch was attached both to these battens and to the rafters themselves.

Thatching—Before taking long straw or combed wheat up on the roof, the thatcher sprays or sprinkles the material with water while it's lying loosely on the ground. This makes the bundles easier to handle. Water reed doesn't require this treatment.

Roofs are begun in essentially the same way with each of the three thatching materials. A bundle of the material is set to split the angle between the eave and the barge, and subsequent bundles are gradually turned so that they are at right angles to the roof's edge at both eave and barge. With threshed long straw, these bundles are called bottles—yealms that are folded lengthwise and tied at the narrow end with twine, tarred cord or twisted straws so that they take the shape of something you might drink claret out of. With combed wheat, they're called wadds—double handfuls of reed pulled from the moistened nitches and tied together near their tops. With water reed, bunches of long reed about 2 ft. in circumference a foot from the butt end are assembled and tied off.

This first course is either stitched to the eaves or held in place with the traditional hazel sways, which themselves are retained by a series of iron hooks driven into the rafters. Sways are hazel rods 4 ft. to 10 ft. long, with a minimum butt diameter of 1 in. They are traditionally lapped by splicing them beneath a single iron hook. Some thatchers use iron sways, but older craftsmen prefer the resilience of hazel.

At the eaveline, a tilting board (drawing, facing page), sometimes called a fillet or arris, is fixed so that it protrudes above the rafter line. It kicks the eave course and the subse-

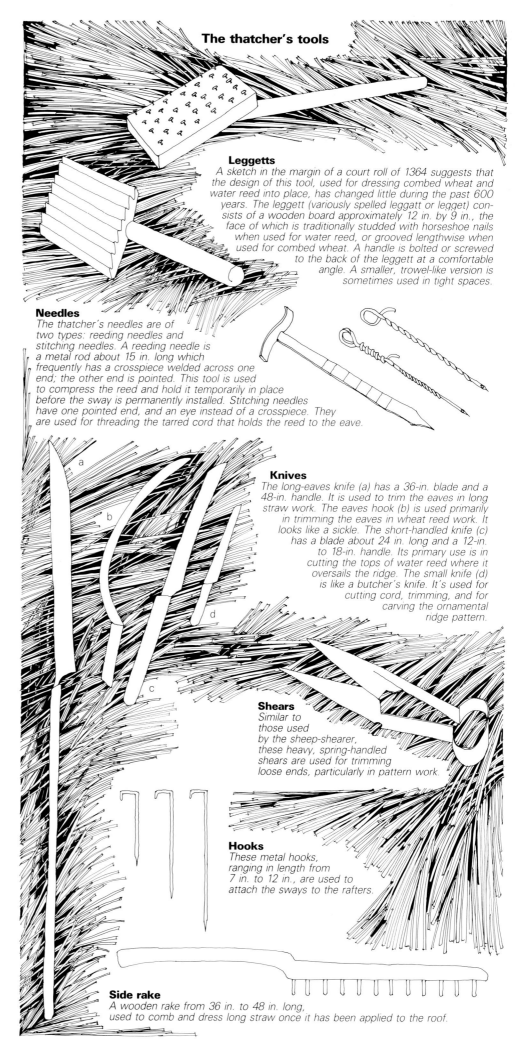

The thatcher's tools

Leggetts
A sketch in the margin of a court roll of 1364 suggests that the design of this tool, used for dressing combed wheat and water reed into place, has changed little during the past 600 years. The leggett (variously spelled leggatt or legget) consists of a wooden board approximately 12 in. by 9 in., the face of which is traditionally studded with horseshoe nails when used for water reed, or grooved lengthwise when used for combed wheat. A handle is bolted or screwed to the back of the leggett at a comfortable angle. A smaller, trowel-like version is sometimes used in tight spaces.

Needles
The thatcher's needles are of two types: reeding needles and stitching needles. A reeding needle is a metal rod about 15 in. long which frequently has a crosspiece welded across one end; the other end is pointed. This tool is used to compress the reed and hold it temporarily in place before the sway is permanently installed. Stitching needles have one pointed end, and an eye instead of a crosspiece. They are used for threading the tarred cord that holds the reed to the eave.

Knives
The long-eaves knife (a) has a 36-in. blade and a 48-in. handle. It is used to trim the eaves in long straw work. The eaves hook (b) is used primarily in trimming the eaves in wheat reed work. It looks like a sickle. The short-handled knife (c) has a blade about 24 in. long and a 12-in. to 18-in. handle. Its primary use is in cutting the tops of water reed where it oversails the ridge. The small knife (d) is like a butcher's knife. It's used for cutting cord, trimming, and for carving the ornamental ridge pattern.

Shears
Similar to those used by the sheep-shearer, these heavy, spring-handled shears are used for trimming loose ends, particularly in pattern work.

Hooks
These metal hooks, ranging in length from 7 in. to 12 in., are used to attach the sways to the rafters.

Side rake
A wooden rake from 36 in. to 48 in. long, used to comb and dress long straw once it has been applied to the roof.

quent brow course upward. This, along with the compression that the hazel sways put on the thatch, places the reed under tension and exposes its ends to the elements. It also gives a slope to the bottom end of the first course that's a little less than the pitch of the rafters and, to a certain extent, to each subsequent course above. The lower ends of each course bristle out (photo center left), resulting in greater durability.

The roof is built up to the ridge by overlapping courses. The thatcher does one section of the roof at a time before relocating his ladder and starting again from the eave. After the eave course, successive yealms, wadds or bunches are temporarily held in place with iron hooks or twisted strands of reed until enough thatch is in place for permanent sways to be installed.

Before beginning each new section of long straw, the thatcher uses a side rake to comb the thatch and give it a dressed appearance by removing any short, loose straw. The traditional tool for dressing both combed wheat and water reed is the leggett, a flat, wooden board about 12 in. by 9 in. (photo facing page). Leggetts have either a grooved surface or one of a variety of facings, most commonly the heads of horseshoe nails or metal strips. Attached to the back of the leggett is a simple handle, the design of which varies from craftsman to craftsman. The leggett is used like a brush to make the reed lie neatly in place and to tighten it beneath the sways.

At the ridge, the procedure varies with the material. With combed wheat, a tightly made roll of straw is tied to the ridge beam with tarred cord (drawing, left). The thatch that oversails the ridge roll may either be cut off or twisted back on itself and forced into the lower portion of the bunch. The knuckles that the thatcher forms this way are held in place by a twisted reed sway sparred into the ridge roll beneath them.

At this stage, a second roll, applied above the twisted knuckles, is "sparred" into place and the pattern course is begun. Spars are 30-in. split hazel or willow rods, pointed at both ends. Each spar is twisted, then bent at its center point, effectively producing a wooden staple whose ends are under tension once it's inserted into the thatch. As shown in the drawing at left, double handfuls of the longest reed are parted down the center, the ends are reversed, and the whole is bent across the angle of the ridge.

Before the top ligger can be installed to secure the pattern course of thatch, the handfuls

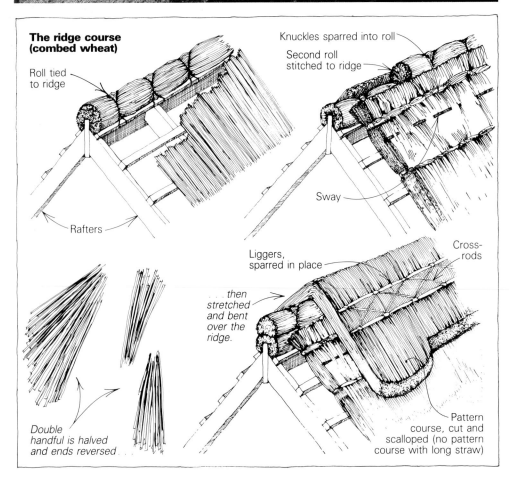

The ridge course (combed wheat)

Knuckles sparred into roll

Second roll stitched to ridge

Roll tied to ridge

Rafters

Sway

Liggers, sparred in place

...then stretched and bent over the ridge.

Cross-rods

Double handful is halved and ends reversed

Pattern course, cut and scalloped (no pattern course with long straw)

of reed must be forced into position and held in place by a combination of needles and reeding pins. The pattern course has to be compressed as tightly as possible.

The course is leveled and secured by liggers, which in turn are held in place by spars. Each craftsman has his own preferred ridge pattern, and by working extra reed into the top course, the additional thickness of thatch necessary for the scalloping is provided. The pattern is cut with a knife and finally trimmed with shears.

With long straw, a 4-in. to 6-in. dia. roll of straw running the length of the roof is secured to the ridge beam by heavy, tarred cord. The tops of the yealms that pass over the apex are twisted together and sparred securely into this roll.

Recall that one end of the yealm is larger than the other. To apply the ridge course, the thatcher straddles the ridge and reverses one half of the yealm, giving him, in effect, an equal-ended yealm. This is then bent across the apex roll. These yealms are pressed tightly together laterally, and the ligger that holds them down is sparred into place. The mark of an accomplished thatcher is the straightness of his line at this point.

The ends of the yealm on each side of the apex are now held in place by two or three more liggers, augmented by cross-rods to provide both additional security and the characteristic ridge pattern.

Sedge *(Cladium mariscus),* an evergreen marsh plant, is used for the ridge cappings of roofs that are thatched with water reed. It is applied over a 4-in. to 6-in. roll of reed tied firmly to the ridge beam. Sedge is traditionally harvested by scythe, and is applied green to the roof. If it dries out before it can be applied, it may be dampened in layers so it can be compressed. Yealms of sedge are bent over the apex of the roof and fastened by spars driven into the reed beneath. All such spars are inserted at an angle below the horizontal so they don't carry water into the thatch. A second, smaller-diameter ridge roll is then sparred through the sedge skirt into the first roll. The next step is to apply another course of sedge yealms. This makes the sedge layer thick enough for the thatcher to cut out his design. Once it has been firmly secured by spars, liggers, and cross-rods, the sedge is scalloped and trimmed with the small knife. The mature, dried sedge shows a serrated, razor-sharp edge.

Speak to the owner of a thatched home and he will extol the silence of the evenings beneath his thatch; the coolness of a summer's midday; the warmth of a winter's afternoon. Not for him the look-alike, angular, mass-produced product in which most of us spend our lives. Not when there's this little thatched place two doors down from the pub. □

Alasdair Wallace is a teacher and writer in Lakefield, Ontario. Photos are by the author, except where noted. For more on traditional British thatching, see The Thatcher's Craft *(£6.50 hardcover, £2.95 paperback, from the Council for Small Industries in Rural Areas, 141 Castle St., Salisbury, Wiltshire, England SP1 3TP).*

Roofing with Sod

Beneath the surface, it's not the same as it used to be

by David Easton

Sod has served as a roofing material throughout history and throughout the world. Sod covered the roofs of 14th-century Pawnee earth lodges, and it made up the walls and roofs of Scotch and Irish folk houses. But sod roofs are most deeply rooted in Scandinavia, where the climate fosters a thick and healthy turf. The traditional Scandinavian sod roof consisted of birchbark sheathing laid across poles running from ridge to wall, with blocks of turf cut from a nearby field and laid on the birchbark. The roofs leaked a bit, to be sure, but some of them lasted for centuries.

As a builder dedicated to the use of natural components, such as rammed-earth walls and soil-cement tile floors, I've been striving to develop a truly dependable sod roofing system. I think a living roof adds immeasurably to the ambience of a home. The first sod roof I built was almost as simple and as cheap as the traditional Scandinavian roof. Covering a small cabin on our ranch in the Sierra Nevada mountains of California, it consists of 30-lb. felt stapled over the roof boards, two layers of 6-mil polyethylene sheeting with asphalt emulsion between them and 4 in. of forest humus seeded with perennial rye. The roofing materials cost about 12 cents a sq. ft., and the labor was all ours. After nine years, the roof still doesn't leak much, but it does leak.

We even raised rabbits on that roof, and our wise old peahen outfoxed the foxes by nesting up there (photo below). The rabbits kept the grass mowed and fertilized, and every so often when one fell off the roof, right into the stew pot he went.

But there are more tangible advantages to a sod roof. Six or eight inches of healthy sod provide a pretty good thermal buffer, helping to keep a house cool in the summer and warm in the winter. One rammed-earth house I built in Calaveras County probably wouldn't

For his sod, Easton uses either lightweight humus sowed with native grasses and wildflowers, or blocks of grass and soil cut from the fields and laid like bricks over the membrane. Left unmowed, Easton's own sod roof makes a hospitable home for his old peahen.

have survived a recent brush fire if not for its sod roof. With the fire fast approaching, the owner mowed his roof and turned on the rooftop sprinklers. The fire swept right past the house, inflicting no damage.

Of course, there are disadvantages to sod roofs, too. They're heavy, calling for stronger and more expensive roof timbers than do conventional roofs. There can also be extra maintenance involved, particularly if the roof is supposed to look well groomed. And if leaks develop, they can be a hassle to find. Unlike conventional roofs, where water runs downhill, 8-in. of soil can create enough hydrostatic pressure to force water upslope and through even the smallest of cracks.

I now have about a dozen sod roofs under my belt, and almost all the problems I've had with them have been due to faulty workmanship along rakes and eaves, and around chimneys, skylights and vent stacks. The hard-learned lesson is that you just can't be too careful, especially when you're burying a waterproof membrane beneath 8 in. of soil. My three most recent sod roofs have weathered for almost four years now, and all are performing as a good roof should. Though there are no guarantees that my sod roofs will last 20 years, I'm encouraged so far. As one who has seen his share of water pass *under* the membrane, I now feel qualified to offer some good tips on roofing with sod.

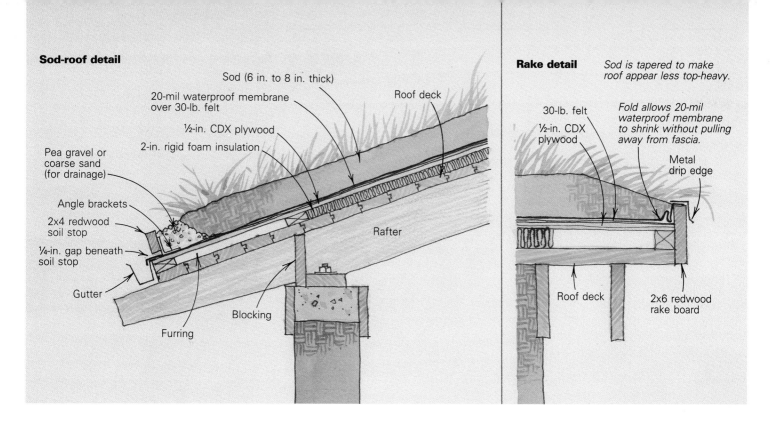

Sod-roof detail

Sod (6 in. to 8 in. thick)

20-mil waterproof membrane over 30-lb. felt

Roof deck

½-in. CDX plywood

2-in. rigid foam insulation

Pea gravel or coarse sand (for drainage)

Angle brackets

2x4 redwood soil stop

¼-in. gap beneath soil stop

Rafter

Gutter

Blocking

Furring

Rake detail *Sod is tapered to make roof appear less top-heavy.*

30-lb. felt

½-in. CDX plywood

Fold allows 20-mil waterproof membrane to shrink without pulling away from fascia.

Metal drip edge

Roof deck

2x6 redwood rake board

Roof-top anatomy—My state-of-the-art sod roof consists of (from bottom to top) the rafters and decking, 2-in. rigid-foam panels, ⅜-in. or ½-in. CDX plywood, a layer of 30-lb. felt, a 20-mil or thicker waterproof membrane, and 6-in. to 8-in. of soil (drawings above). I substitute furring strips for the insulation on the overhangs. A narrow band of gravel or sand promotes drainage along the eaves and around chimneys and skylights.

Every sod roof my crew and I have built sits on top of an open-beam gable structure—heavy timbers decked with 1½-in. T&G decking. But trusses or 2x rafters will work just as well, provided the roof is engineered to support the 100 lb. to 140 lb. per sq. ft. that 8 in. of soaking wet soil will weigh (if you live in snow country, you have to figure snow loads into the equation, too). As long as the roof framing can handle the weight, the building departments I've dealt with have had no problem with the use of sod as a roofing material.

Though traditional Scandinavian sod roofs are typically steep, we stick to a 6-in-12 pitch or under for our roofs (we prefer a 4-in-12 pitch). Sod bricks allow a steeper roof, but loose soil will creep downslope in a heavy downpour, at least until the grass roots have had a chance to knit the turf together. There's no real rule-of-thumb regarding roof types. I've seen sod installed over everything from flat roofs to complicated compound-curve roofs. If the sod will stay put on the roof, it will do the job.

A top-of-the-line sod roof isn't cheap. By the time you add up the cost of insulation, sheathing and waterproof membrane and factor in the labor, the total cost approaches that of a cedar-shingle roof. Factor in the hefty roof framing required (which we use anyway for aesthetic reasons), and a sod roof can cost double that of a shingle roof.

Insulation and sheathing—Once the roof is decked, the insulation is the first component of the sod roof to land on it (top photo, next page). Our early sod roofs had no insulation other than that provided by the sod itself. But I'm convinced that insulation would have been worth the cost, especially during cold, wet weather. We use Thermax polyurethane insulation (Celotex Corp., P. O. Box 31602, Tampa, Fla. 33631) because polyurethane packs a lot of R-value into a small amount of space (R-16 for 2-in. thick panels). There is no need to insulate overhangs, so at the eaves and rakes we nail a 2-in. thick grid of scrap lumber to the deck, producing a flat nailing surface for the plywood sheathing. It takes longer to install this furring grid than to lay down rigid insulation instead, but it saves the cost of a few sheets of Thermax.

We lay the insulation from the ridge down, beveling the top edges of the ridge pieces for a snug fit. Each sheet is placed tightly against its neighbor and tacked down with a couple of 12d nails. We let the bottom course lap the eaves where it will and then fill in the leftover space with the furring grid.

Once the insulation and furring are in place, we sheathe the whole roof with plywood. Here we use 16d galvanized nails, which are long enough to pass through the plywood and insulation and to penetrate more than an inch into the roof deck. This time we work our way from the bottom of the roof to the top in order to offset the horizontal seams of the plywood and the insulation board (we also offset the vertical joints).

Not all our sod roofs have gutters, but when they do, we install them after the plywood is in place. We use 5-in. deep galvanized gutters, bending the backsides over at a right angle to form a 1-in. lip. The lip is then nailed securely to the edge of the roof with 16d galvanized nails. Bending the gutter

like this shortens the effective depth of the trough, but it makes for a positive connection to the roof and allows us to wrap the waterproof membrane over the lip and into the gutter. Also, this raises the gutter so that its bottom edge is flush with or above the underside of the decking, allowing us to extend the rafter tails out beyond the gutters. If you like the look of exposed rafter tails as I do, this bend in the gutter is essential.

The drawback to bending the gutter is in dealing with that day in the distant future when the gutter needs to be replaced. Someone will have to lift up the membrane, yank out the old gutter and spike in a new one. If the budget allows you to use copper gutters, you won't have to leave instructions for installing a new gutter in your will.

Next come the rake boards, which retain the soil along the rakes. We use redwood 2x6s or 2x8s for this, but pressure-treated lumber works just as well. A 2x6 doesn't retain as much sod as a 2x8 does, but I prefer to use the slimmer boards and to taper the sod near the edges so the roof appears a little less top-heavy. We spike the rake boards to the ends of the decking with 20d galvanized spikes, aligning their bottom edges with the bottom of the roof decking.

Waterproofing—Before we install the waterproof membrane, we staple 30-lb. felt over the sheathing to protect the membrane from nail heads and wood splinters (bottom photo, next page). Courses lap a minimum of 3 in., and each course is cut long enough to turn up the sides of the rake boards (we're careful here to bend the felt at tight right angles). If there's a gutter, we lap the bottom course of felt over the bent section of the gutter and glue the felt to the gutter with mastic.

For the membrane itself, we use 20-mil chlorinated-polyethylene sheeting called

NobleSeal 220 (The Noble Company, 614 Monroe St., Grand Haven, Mich. 49417). Available in 5-ft. by 100-ft. rolls, it's commonly used for pond liners and for waterproofing under concrete slabs.

To install the membrane, we work our way from the eaves to the ridge (bottom left photo, facing page). Adjacent courses are lapped 4 in. to 6 in. and chemically fused with a brush-on sealant called Nobleweld (also available from The Noble Company), which permanently bonds the seams and prevents water and root penetration. We're extra careful with this part of the job because roots will penetrate even the smallest of openings in search of water and warmth. We position the first course so that the bottom edge overlaps the felt and either runs down the back of the gutter about two inches or, if there is no gutter, bends over the edge of the roof. Each course laps the tops of the rake boards on both ends, and the top course crests the ridge and drops down the opposite side of the roof. We staple the membrane only at the ends and along the top of each course, where the staples will be covered by the next course up.

Though the manufacturer claims that NobleSeal shrinks less than other membranes, a tightly stretched membrane will eventually shrink enough to pull away from the rake boards and cause leaks (naturally, we learned that the hard way). So now we cut the membrane for each course a few inches long and tuck an extra fold into it alongside the rake at both ends (top right drawing, previous page). As the membrane shrinks, it takes up this slack.

Vents, skylights and chimneys—Roof penetrations present special problems with sod roofs. In fact, we do our best to keep them to a minimum. For vent stacks, we use standard metal flanges with neoprene gaskets, which we nail on top of the felt, and cut round holes in the membrane to fit over the flanges. As an extra precaution, we fit a second piece of NobleSeal (about 2 ft. square) over the vent stack and glue it securely to the primary membrane with Nobleweld. We cut the hole in this piece slightly undersize so that the membrane stretches tightly around the flange. We use this same approach for stovepipes.

Because sod is substantially thicker than most other roof coverings, most skylights require a higher curb than usual to elevate them above the plane of the roof. We usually build a 2x12 redwood frame that slips down into the rough opening and projects about 7½ in. above the plywood sheathing, serving as both the finish trim on the interior and the curb on the roof. This curb requires careful and systematic flashing with four 12-in. wide strips of 30-lb. felt and four 18-in. wide strips of aluminum or galvanized-steel flashing (drawing, facing page).

Flashing begins right after we install the 30-lb. felt over the entire roof, with the felt cut to fit snugly around the 2x12 frame. The

felt flashing is installed first. We start by cutting the first piece 12 in. longer than the width of the frame, folding it in half lengthwise and installing it against the bottom of the frame. At the corners of the frame, we carefully slice halfway through the flashing and wrap the ends around the corners, applying a healthy dab of mastic to prevent leaks. The two side pieces are cut, folded and installed the same way, with their bottoms lapping the ends of the bottom flashing. Again, we use a dab of mastic to prevent leakage. The top flashing matches the first piece and laps the tops of the side flashing.

Next, we install the waterproof membrane over the entire roof, cutting it for a close fit against the skylight frame, without turning up the sides (turning the membrane up prevents the final layer of flashing from lying flat). What we have at this stage is a roof covered in its entirety with both felt and the waterproof membrane, and the skylight curb flashed with strips of 30-lb. felt. The final step is for us to flash the skylight frame with the 18-in. wide strips of galvanized metal, using the same careful method described above. The bottom strip sits over the membrane, as do the side pieces, until they reach the top corners of the curb. Here, they tuck under the membrane through a neatly cut horizontal slit in the membrane. The top flashing also slips under the membrane and laps over the top corners of the side flashing.

Once again, we apply a generous dose of mastic where it's needed. This completes the primary flashing. Depending on how far down over the curb the skylight will sit, we sometimes install counterflashing—a 6-in. wide length of galvanized metal that's nailed

Sod-roof construction. The roof structure for a sod roof has to be sturdy enough to support 6 in. to 8 in. of soaking wet soil (100 lb. to 140 lb. per sq. ft.). Once the roof is framed and decked, 2-in. thick polyurethane insulation is installed over the deck (top left photo). To save on insulation, Easton furs the overhangs with 2-in. thick scrap lumber. In the photo below left, the insulation is topped with CDX plywood, and 30-lb. felt is being stapled to the plywood to protect the waterproof membrane from plywood splinters and nail heads. The membrane is a 20-mil chlorinated-polyethylene sheeting commonly used for waterproofing concrete slabs. In the photo below, the membrane is being laid directly over the roofing felt, with adjacent courses lapped 4 in. to 6 in. and fused together with a brush-on glue. The ends of each course wrap over the tops of the rake boards and will be covered by metal drip-edges. To promote drainage, a narrow band of coarse sand is placed along the eaves (photo right) and around chimneys and skylights. Sand and sod are contained at the eaves by a redwood 2x4 fastened to the roof with galvanized angle brackets and screws. Because summers are dry in California, Easton typically installs rooftop sprinklers or drip-irrigation systems to keep the grass green and the house cool.

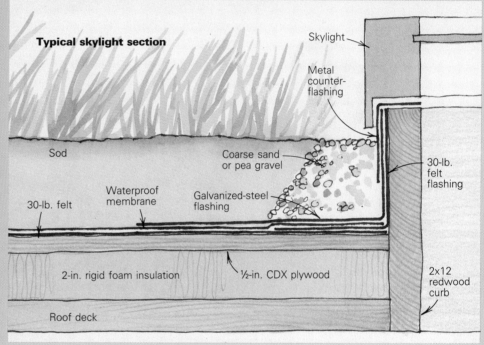

Typical skylight section

Skylight

Metal counter-flashing

Sod

Coarse sand or pea gravel

30-lb. felt flashing

Waterproof membrane

30-lb. felt

Galvanized-steel flashing

2-in. rigid foam insulation

½-in. CDX plywood

2x12 redwood curb

Roof deck

---

to the top of the curb and bent down over the primary flashing.

Masonry chimneys are flashed the same way (the chimney is treated as a curb), except that they require the usual counterflashing embedded in the mortar joints.

Edge treatment—Once the roof is wrapped with the waterproof membrane, three steps remain before we start hauling up the topsoil. First, we fasten a 2x2 metal drip edge to the rake boards with 5d or 6d nails. This prevents water from working its way under the membrane and protects the edge of the membrane from the sun. Second, we install a redwood 2x4 on the roof along the eaves to retain the soil (top photo), holding the 2x4 ¼ in. above the membrane to allow drainage. The redwood is held fast with galvanized-metal angle brackets screwed to the roof,

with big gobs of silicone caulk covering both the brackets and the screws. Finally, we place a narrow band of coarse sand or pea gravel along the edge of the 2x4 and around skylights and chimneys to promote quick runoff.

Topping it off—For soil, we use the best quality, lightest humus (loose, friable organic soil) we can gather at the building site, or we use the old Scandinavian trick of digging sod blocks and carrying up the grass and the soil together (we've tried this one a couple of times, but always have trouble finding grass thick enough to stay in one piece during transport).

Whichever method we use, getting the soil up on the roof is a chore. We've used conveyers, tractors, a backhoe, ramps and wheelbarrows, and even 5-gal. buckets. If we use loose soil, we rake it on the roof until it's

smooth; then we sow it with native grasses and wildflowers. Sometimes we plant a few bulbs in it, too.

What about watering and mowing? Here in California, where the summers are dry, we always install sprinklers or drip irrigation systems on the roof. In addition to keeping the roof green and healthy, the water makes the roof work as an evaporative cooler, reducing the temperature in and around the house.

Most people I know who have sod roofs don't mow them. Rather, they let them grow wild and shaggy, building up humus and root mass over the years for increasingly better insulation (top photo, p. 80). Given enough time, the roof will look like it grew there. □

David Easton is a contractor and earthbuilder in Wilseyville, California. Photos by the author.

The Cotswold Slat Roof
Reviving the age-old craft of shingling with stone

by Alasdair G. B. Wallace

Slat roofs are one of the most striking features of the Cotswold villages. Roofs and walls come from locally quarried limestone.

The Cotswold villages, nestled into the sheltered valleys of the Coln and Windrush Rivers, are among the prettiest in England. The limestone cottages, barns and churches of the villages have aged over the centuries to a rich golden hue and their steeply pitched roofs are striking. Nearly every roof is made of irregularly shaped stone shingles, or slats (photo above), which are riven from locally quarried limestone. Unlike a standard shingle roof, however, the exposure of the slats decreases incrementally from the eave to the ridge. Also, the slats are longer, wider and thicker at the eaves than they are at the ridge. The effect of a Cotswold slat roof has been likened to the orderly beauty of a fish's scales or a bird's feathers.

Concerns for the preservation of the Cotswold villages have led to a revival of the ancient craft of slatting. Cotswold slat roofs are labor-intensive and expensive. The roof slatting of one large manor I worked on during a visit to England took two skilled men four months to complete. They gathered slats from three old barns, and much of their time was spent in sorting, hauling and trimming them. Many Cotswold roofs have lasted from 200 to 400 years with little maintenance, so the cost, when amortized over the years, may be appreciably less than for a conventional roof.

Quarrying the pendle—The Romans appear to have been the first to mine stone slats in the Cotswolds due to the region's lack of lumber for wood shingles. Traditionally, the *pendle* (a block of stone consisting of thin layers separated by fine films of clay) was quarried between Michaelmas (September 29) and Christmas. Mines could be up to 60-feet deep. Once quarried, the villagers covered the pendle with sod to retain the earth's moisture, or *sap*. At the first heavy frost, church bells were rung, and the villagers hastened to remove the

sod. A week-long frost, which usually occurred in January or February, was enough to split the layers of stone.

Once the pendle was split, slatters worked until the following autumn to convert it into finished slats. Working quickly, a slatter would shape three edges with his slatting hammer and his *crapping stone*, a stone equivalent to the blacksmith's anvil (photo, facing page). Then he finished the edges of the slats by *battering* them with the side of his hammer. Two-hundred-and-fifty slats were considered a day's work.

As each slat was finished, the slatter made a single hole in the head of each by tapping lightly on the back side of the slat with a pick. The slat pick consisted of a short ash handle with a square-sectioned steel point wedged through one end. Tapping the point on the underside of the slat broke out a small piece on the face, leaving a small crater. The point was then inserted into the rough hole and twisted to produce the finished round hole. The fine art of chipping a hole without breaking the slat has been all but lost, the electric drill and masonry bit having replaced the slatter's pick on the job site.

Finding the presents—Today, most riven slats come from old buildings, but a few builders in the Cotswold area have reopened the old pits and quarries. Instead of relying solely on the pendle, they occasionally carve out a pasture with heavy equipment, scooping up layers of stone lying no more than 4 ft. below the sod. This surface layer is traditionally called the *presents*, the gift of thousands of years of frosts. The presents are the hardest stone of the quarry, and they yield a superior roofing material. Their surface texture is much rougher than that of the pendle and offers a more hospitable surface to moss, lichen and stone-crop which, over the years, add their own inimitable patina to the Cotswold roofs. The disadvantage is their inconsistent color, which ranges from ochre through buff to an amber chestnut. Another drawback is that they lie beneath valuable pasture land, which is why they were rarely used in the past.

Today, presents are sawn roughly to the size of slats with a circular saw, then are dressed by hand as were traditional slats. Great care is taken that the slat edge not be even in thickness. Irregularities encourage interlocking of the slats once they're on the roof, and this helps to keep out rain and hold the slats in place against the wind. Slats vary in thickness from ½ in. to 1 in.

Once the slats have been dressed, they are sorted and transported by truck to the building site, where they are stacked according to size. The number of slats of each size required for a specific roof is calculated at the quarry, using a measured drawing of the roof and a row count furnished by the slatter.

Supporting the slats—Slat roofs are incredibly heavy, weighing about 2,200 lb. per square (asphalt shingles typically weigh about 200 to 300 lb. per square). The roof pitch is usually at least 12-in-12 to encourage drainage. This is important because slats are slightly porous. Some large barns were purposely built with a concave ridge, which minimized the effect of the wind and ensured a tighter bond between slats.

The bottom of a slatted roof tilts upward in order to throw rain clear of the walls. This is usually accomplished by one of two methods. The first method requires the use of a *cocking piece*—a heavy board two to three times the thickness of a regular batten—affixed along the bottom of the rafters. The second method positions the tail of each rafter 4 in. in from the exterior face of the wall so that the top of the wall lifts the slats upward (drawing at right).

Traditionally, the slats were laid on a bed of moss, hay or straw, which was provided and installed by the *mossier*. The mossier

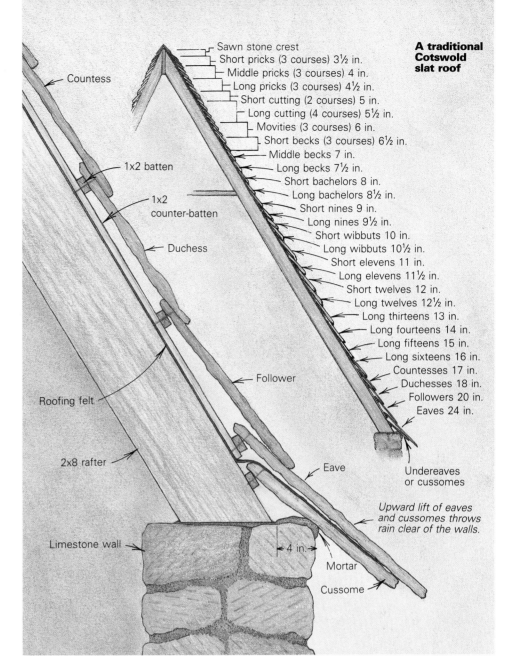

A traditional Cotswold slat roof

- Sawn stone crest
- Short pricks (3 courses) 3½ in.
- Middle pricks (3 courses) 4 in.
- Long pricks (3 courses) 4½ in.
- Short cutting (2 courses) 5 in.
- Long cutting (4 courses) 5½ in.
- Movities (3 courses) 6 in.
- Short becks (3 courses) 6½ in.
- Middle becks 7 in.
- Long becks 7½ in.
- Short bachelors 8 in.
- Long bachelors 8½ in.
- Short nines 9 in.
- Long nines 9½ in.
- Short wibbuts 10 in.
- Long wibbuts 10½ in.
- Short elevens 11 in.
- Long elevens 11½ in.
- Short twelves 12 in.
- Long twelves 12½ in.
- Long thirteens 13 in.
- Long fourteens 14 in.
- Long fifteens 15 in.
- Long sixteens 16 in.
- Countesses 17 in.
- Duchesses 18 in.
- Followers 20 in.
- Eaves 24 in.

Countess

1x2 batten

1x2 counter-batten

Duchess

Follower

Roofing felt

2x8 rafter

Eave

Limestone wall

4 in.

Mortar

Cussome

Undereaves or cussomes

Upward lift of eaves and cussomes throws rain clear of the walls.

Once the slats are split from the quarry stone, the slatter trims them to size with a slatting hammer. The edges of the slats are then finished by battering them with the side of the hammer. A crapping stone provides support for the work.

Swept valleys were once the mark of the accomplished slatter. The process, called valleying, required no nails or pegs. Triangularly shaped slats were woven together, and their weight held them tightly in place. Valleying appears to be a lost art.

After the last row of slats has been laid on the roof, the ridge is sealed with the traditional limestone crest. The crest is firmly embedded in mortar to seal the ridge against the weather.

also filled the gaps between shingles with the same material in order to keep out driven snow. Today, moss has been replaced by a heavy roofing felt, which is laid directly on top of the rafters.

Wood counter-battening is sometimes nailed along the top edge of each rafter over the roofing felt. The battens that support the slats are then installed perpendicular to the counter-battens. This system encourages air circulation between the roofing felt and the slats. The 1x2 rough-sawn battens are secured with 2½-in. galvanized nails.

Cutting a whippet stick—The exposure of the slats decreases as courses approach the ridge, so the battens have to be spaced accordingly. Though contemporary slatters frequently use a tape measure to determine the spacing of battens, traditional slatters used a homemade slatter's rule, or *whippet stick*. A whippet stick was a straight-grained piece of hardwood approximately 1 in. by 1 in. in section and 30-in. long. A series of horizontal marks were scored on the stick, and a nail protruded about 1 in. through one end of it. Slatters could quickly determine the position of each batten on the roof by butting the stick against the top of the preceding batten and aligning the new batten with the appropriate mark on the stick. The whippet stick also allowed slatters to measure the length of each slat to establish its location on the roof. To do this, they inserted the nail on the stick into the hole in the head of the slat and read the length of the slat against the horizontal score marks. Each whippet stick was made by a slatter to suit his own preference and to match most nearly the materials supplied to him by the local quarry.

The Romans affixed their slats with iron nails, but later slatters used oak pegs. They drove them through the holes in the heads of the slats. The countersinking in the face of the slats housed the heads of the pegs. The slats were then hung over the oak battens.

Slatting the roof—Traditionally, each course of slats had its own name, which varied from one community to another (drawing, previous page). Each slat was ½ in. longer than the one above it. With very few exceptions, today's slatter has discarded the old names in favor of a series of numbers. The slatters I worked with used slats ranging from 6 in. to 24 in. long (from top to bottom), instead of the traditional 3½ in. to 24 in. long. From 6 in. to 12 in., they reduced the exposure of each course by ½ in. From 12 in. to 24 in., though, they reduced the exposure by a full inch.

The undereaves, or *cussomes*, are the first slats to go on the roof. They project beyond the wall by about 6 to 8 in., and slatters slip them face down beneath the roofing felt because they are visible from below. Any moisture that penetrates beneath subsequent slats is thus shed away from the building. The cussomes are bedded in mortar, which ensures a tight bond between the roof and the wall. As with the ensuing courses, the cussomes are fastened to the batten with galvanized nails.

The eaves course is laid face up on top of the felt and the cussomes. This course bears the brunt of the weather, so it is considered the most important course on the roof. The slatter reserves his largest, soundest slats for this location. Slats here are up to 40-in. wide and their length projects beyond the cussomes by about 3 in., which keeps the eaves dry. Extra care is taken to offset the joints from those in the cussomes by at least 3 in., and the joints should be offset by a minimum of 2 in. on the rest of the roof. Also, the bottom of each shingle should lap the nail which affixes the second course beneath it by at least 2 in. This and the side-lap will ensure a tight roof.

Using a mason's line to keep the courses straight, slatters work their way up the roof. Slats are not of a uniform thickness, so the slatter must constantly be aware of any disparity in thickness. An area of thicker slats in one course requires that thinner slats be located above it to compensate. In this manner the roof will progress evenly without any ripply effect. Minor variations are adjusted at the top edge of each slat. The width of the battens will accommodate variations up to ½ in.

While some of the larger slats have two nail holes, the majority are fastened with one nail only. Any lateral swinging movement is impeded by the adjacent slats. One advantage of this technique is that it makes repairs easier to handle. If a slat needs to be replaced, adjacent slats can be lifted by hand and swung out of the way.

Along the verge of the roof (the edge along the rake), the slatters embed the slats in mortar. This added thickness creates a slight upturn that encourages water to run down the center of the roof rather than over the verge.

In years long since passed, the mark of a finely executed roof was the swept valley, whereby one roof plane melded gently into another (photo facing page, top). This process, called *valleying*, required no pegging or nail-ing. Small triangular slats were woven and dovetailed together. Their weight wedged them tightly together on the steeply pitched roofs and held them in place. Valleying, however, appears to be a lost art. These days slats are carefully mitered and laid over lead flashing (photo below). Individual shingles are marked on the roof, then cut on the ground with a circular saw fitted with a masonry blade.

When the last course of slats has been laid on the roof, the ridge is sealed with a traditional sawn-stone crest, which is firmly bedded in mortar (photo facing page, bottom).

It's encouraging to know that the traditional craft of slatting is still practiced. Visit a Cotswold village one harvest evening when the golden stubble finds its echo in the mellowed stone roofs, and you'll understand that the tradition of stone slatting must not be permitted to slip into oblivion. □

Alasdair G. B. Wallace, of Lakefield, Ontario, is a contributing editor to Fine Homebuilding. *Photos by the author.*

Manufactured slates

The increasing demand for both new and used slats has prompted the development of reconstructed stone slates. One of the manufacturers is E. H. Bradley Building Products Limited of Swindon, a company that introduced its Bradstone Cotswold Slates in 1970. The slates are available in the full range of diminishing sizes and widths, and they're manufactured in one of two shades—weathered buff, which most nearly matches the natural Cotswold shade, or grey/green, which resembles the natural stone of Scotland and Wales. The color pigments are blended throughout the slate. Despite their varying widths and careful color-gradations, though, manufactured slates are discernible from quarried slats by the regularity of their appearance on the roof (photo below).

Each predrilled slate has a code number cast into its upper face, where it will be covered as part of the headlap (the part of a slate that is lapped by the slate above it). This number indicates not only the length of the slate, but its width. To further facilitate installation, slates may be purchased with cast nibs in addition to, or in place of, the drilled holes. The nib is a ridge on the underside of the slate, which is slightly above the two nail holes and which spans the gap between them. The sole function of the nib, like that of the traditional oak peg, is to enable slatters to hang the slates over the battens. The weight of each slate holds it in place. Standard procedure is to nail every fifth course. Bradley also supplies a full range of wings and slips. With them, roofers can emulate the traditional swept valleys.

Manufacturer's tests have shown that these slates, which average ⅝-in. in thickness, can be used on roof pitches down to 30° (about 7-in-12). Installed at the recommended minimum 3-in. headlap, a roof of Bradstone slates will weigh about one-third as much as a roof made of natural slats.
—A. G. B. W.

Manufactured slates on the new roof (shown at right) are more uniform in appearance than the quarried limestone slats to the left. The numbers stamped on top of the slates indicate their position on the roof. Slates can either be hung over the battens on cast nibs or nailed to them. The mitered corners are installed over lead flashing.

Metal Roofing

Sorting through the confusing array of metals, patterns and coatings

by J. Azevedo

Mention metal roofing, and most people will conjure up an image of rusty, corrugated sheets nailed to a barn roof, with seams peeled back by the wind (photo below). The galvanized panels that roofed country houses and outbuildings around the turn of this century did not weather gracefully, and so established a strong prejudice against metal roofing. But times have changed. Today, architects and builders are discovering that modern metal roofing panels bear little resemblance to the old "tin roofs." Roofing manufacturers now offer a variety of metals, patterns and coatings in a wide range of colors. New panel patterns provide a tight weatherproof skin with a clean, bold style that suits contemporary houses.

In many ways, metal is the ideal roof covering. Metals are strong yet lighter than other coverings, and rigid but not brittle like slate, tile and fiberglass shingles. Metal roofs won't burn, and only the most intense hail will damage them. Pitched metal roofs shed ice, snow and rainwater without eroding. Large sheets of metal roofing typically present few joints and crevices where water might work an insidious path toward the rafters.

On the negative side, metal expands and contracts noticeably with changes in temperature. Metal roofs must be designed and installed to allow for these dimensional changes. Also, metals corrode, some faster than others (see the sidebar on p. 92), and susceptible metals must be protected. Finally, residential metal roofs aren't cheap. They cost at least as much as a premium composition-shingle roof, or many times more.

The sound of rain on a metal roof has been argued as a plus and a minus, with extremists on both sides. Nostalgic recollections of the gentle ring of rain on a porch roof might be viewed skeptically by a person suffering beneath the deafening downpour of a thunderstorm on an open aluminum-on-purlin shed roof. However, if you put on a metal roof, you probably will not hear either of those sounds. With metal on solid decking over code insulation and interior covering, even the hardest rains will produce but a faint tintinnabulation.

The decision to try metal roofing is only the beginning. A vigorous, competitive market in metal roofing has produced a bewildering array of metals, styles, patterns and protective coatings. Your choice of the combination of metal/pattern/coating will determine the roof's initial cost, future maintenance requirements and long-term durability.

Metals—Probably every malleable metal has at one time or another been used as roofing, but only a few have proven especially suitable. Selecting one is a matter of weighing durability and cost. As with most materials in a supply-and-demand economy, the metals that last longest cost the most initially. The decision also involves aesthetic judgments.

Copper is historically the roofing metal of choice—the standard for judging all others. At least one public building in every city has a copper roof, perhaps a dome with green patina, recalling a time when roofs were designed to last the life of a building, and buildings were built to last for generations. Copper's longevity stems from two things. First, it is relatively inert, and so does not corrode rapidly when exposed to the elements. Second, the brown-green patina, which is actually a thin layer of surface corrosion, acts as a skin that protects the metal from further degradation. Among its other attributes, copper is easily formed and a joy to work.

On the negative side, copper will corrode other metals it touches, and drainage from a copper roof will stain paint. Lime and other alkalis will stain copper black. While not the most expensive roofing metal, copper is right up there near the top (see the chart below). Because copper can last hundreds of years, its expense might be justified if the building is designed to survive as long as the roof. For residences, however, copper roofing is specified for reasons other than economy.

Stainless steel, the most expensive roofing steel, incorporates chromium and nickel into the alloy. These two ingredients give stainless its durability and resistance to rust and corrosion. Typical roofing stainless, known generically as dead soft type 304 or 18/8, contains 18% chromium and 8% nickel. The conventional annealed finish is semi-bright. A "rough-rolled" finish gives a softer, less reflective look. But both finishes stay bright—

Metals comparison chart				
Metal	Thickness (in.)	Weight[1] (lb./sq.)	Expansion[2] (in.)	Relative Cost[3]
Zinc alloy[4]	.027	125	1.42-1.54	135
Copper (16 oz.)	.022	125	1.12	100
Terne-coated stainless	.015	89	1.06	93
Stainless 304	.018	79	1.15	61
Terne (28 ga./ 40 lb.)	.015	90	.83	42
Aluminum	.032	58	1.54	42
Cor-Ten A	.048	270	NA	29
Galvanized steel	.022	113	.78	18[5]

[1] *Based on the coverage of a standing-seam pattern.*
[2] *Potential free movement of a 100-ft. metal sheet with a temperature change of 100°F.*
[3] *Cost of metal sheets as a percentage of the cost of copper ($2.10/sq. ft., small quantities, Jan. 1984, on West Coast).*
[4] *Microzinc 80 (W. P. Hickman Co., P.O. Box 15005, Asheville, N. C. 28813).*
[5] *Prepainted galvanized sheet costs about 30% more.*

some would say too bright—through years of exposure. Ironically, this resistance to tarnishing probably keeps this otherwise fine roofing metal from being specified more often.

When stainless eventually does corrode, the damage comes as localized pits. A small pinpoint of corrosion eats down into and through the metal, expanding as a deep vertical shaft. This type of corrosion can cause leaks in a roof that looks sound at first glance. Pitting is most severe in coastal areas because salt in the air accelerates the process. Despite this problem with pitting, stainless resists rust and corrosion better than most metals. In addition, stainless can be mated with most other metals without danger of galvanic action. It costs about two-thirds as much as copper.

Cor-Ten is the trade name for the "weathering" steel made by United States Steel (600 Grant St., Pittsburgh, Pa. 15230). Rather than resisting rust, Cor-Ten welcomes it. When it's exposed to the weather, Cor-Ten develops a thick scale of oxides, which armor it against deep corrosion. Cor-Ten (photo top right) uses a special alloy (carbon/manganese steel with small amounts of silicon, phosphorus, chromium, copper and nickel) to fend off deep rusting. For this special weathering process to work, Cor-Ten roofing must be thicker than other standard roofing panels. Commonly specified in 18 ga., or .048 in. (more than twice as thick as regular galvanized roofing), a Cor-Ten roof weighs about the same as a standard asphalt-shingle roof. Besides weight, the other disadvantage of Cor-Ten is that its runoff is messy. The soft red hue that looks natural and attractive on the roof has less appeal when it drips down a painted rake or stains the groundcover around the downspouts.

Galvanized and aluminized steel. Regular mild steel neither resists corrosion like stainless nor tolerates it like Cor-Ten. Therefore, ordinary steel used for roofing is always plated with some other metal that resists oxidation and corrosion. The zinc plating of hot-dipped galvanized sheet is the most common treatment, but hot-dipped aluminum plating (aluminized steel) and aluminum/zinc alloy platings called Galvalume (Bethlehem Steel

Then and now. **Hastily constructed buildings covered with unpainted galvanized panels, such as the wind-blown barn on the facing page, set a tone that soured this country's perception of metal roofing. But today, there are sleek metal roofs with a sophisticated array of patterns, colors and finishes, like the terne-coated stainless-steel roof on the Hult Center for the Performing Arts in Eugene, Ore., right. Cor-Ten steel, top, is designed to develop a thick surface layer of rust that will protect the metal from deep corrosion. The material develops a pleasing color, but care has to be taken to divert messy runoff away from walls and landscaping. By far the most popular metal roofing is painted galvanized steel, far right. This house is topped with a typical V-rib steel roof, which is secured with exposed fasteners. This particular brand is called Amer-X6. It comes pre-painted from the factory (Engineered Components, Inc., P.O. Box 9, Tualatin, Ore. 97062), with a 20-year guarantee on the integrity of the paint film.**

Photos this page: J. Azevedo

On-site forming. **The portable roll-forming machine, top, is a revolutionary development in metal roofing. With this machine, a roofing contractor can call out exact dimensions to the operator, cutting down on the mistakes that can occur if precut material has to be ordered. This can be especially helpful on complicated roofs. Once the pans are in place, their standing seams can be crimped by hand or by a seam-bender on wheels, above.**

Corp., Bethlehem, Pa. 18016) are also available. They have a longer lifespan, especially in coastal environments where the salt air would quickly eat away zinc plating.

Galvanized-steel roofing is the least expensive, and not surprisingly, the most popular type of metal roof. Strong and light, galvanized roofing comes in a wide variety of patterns and colors. Galvanized panels are usually painted. Left unpainted, the zinc plating eventually wears away, exposing bare steel, which soon rusts. Regular painting and maintenance will substantially extend the life of the panels.

In contrast to other roofing metals, galvanized steel comes in a broad range of quality. Thickness ranges from 30 ga. (.015 in.) to 20 ga. (.036 in.), and even the definition of "gauge" varies. Some metals are judged by a standard that doesn't apply to another, so it's best to ask for thickness in decimals. The galvanized plating may be either G90 (1.25 oz. of zinc per sq. ft.) or G60 (.83 oz. of zinc per sq. ft.). More zinc is better. Builders shopping for a steel roof need to be especially careful about checking quality when comparing prices.

Terne metal, an alloy of lead and tin on mild steel, has a long history in roofing. It was first imported to this country from Wales in the early 19th century. The steel base gives high strength and low expansion while the terne plating offers resistance to weathering. Terne metal cannot be left bare, however, because the plating is porous. Left exposed, terne metal will rust from within. The manufacturer of terne metal (Follansbee Steel Corp., Follansbee, W. Va. 26037) recommends priming terne with red iron oxide in linseed oil and repainting every eight to ten years.

Rusting is not a problem with terne-coated stainless (photo previous page, bottom left). It's a recent product from Follansbee that allows the terne coating to weather to its natural dark grey (the word *terne* means dull, as distinguished from bright tin plate). Terne alloy is more reactive than stainless, so the terne plating will weather away before the stainless core begins to corrode. The terne/stainless combination can be considered a permanent roofing metal in a class with copper, with a price tag to match.

Zinc. Rolled zinc roofing, quite common in Europe, has never been popular in this country. Although zinc weathers to an attractive natural grey and has a life expectancy of more than 100 years in most environments, the metal does have some idiosyncracies that discourage its use as a roof covering. Zinc can gradually yield and stretch (creep) under load. To combat this tendency, the W. P. Hickman Co. (P.O. Box 15005, Asheville, N. C. 28813), the major manufacturer of zinc products, alloys copper and titanium with zinc. These zinc alloys, marketed under the trade name Microzinc, are more resistant to sagging than pure zinc. Still, zinc-alloy roofing is better installed on a solid deck rather than over purlins.

Zinc alloys, together with aluminum, have the highest expansion coefficients of any of the common roofing metals. For a given change in temperature, zinc alloys will expand (or contract) twice as much as steel. The roofing pattern should allow for this. Also, when forming zinc-alloy sheets, shop temperatures must be kept above 50°F or the metal can crack on the convex side of the bend.

Zinc corrodes easily by galvanic action and so must be isolated from most other metals. Furthermore, the natural acids in oak, redwood and cedar will attack zinc. A coating of asphaltum paint can provide the necessary separation between zinc panels and wood or other metals. When zinc alloys are protected from galvanic or acid corrosion, the metal weathers slowly. The surface develops a uniform carbonate coating, which, like the patina on copper, protects the metal against further corrosion. Runoff from zinc-alloy roofing will not stain adjacent surfaces. All things considered, zinc alloys would be an attractive roofing material if they were not so expensive. Several years ago, Microzinc cost less than half as much as copper, but now it costs at least a third more. Zinc's few nasty habits, which could be overlooked when the price was low, seem glaring in light of current costs.

Aluminum. Pure aluminum is soft and weak, but alloys of aluminum and other metals are much stronger and still light. Typical alloys used for roofing include aluminum/manganese 3003 and 3004 and, less commonly, aluminum/magnesium 5005 (the numbers are standard designations that refer to the composition of the alloy). Aluminum roofing panels resist corrosion and tolerate coastal salt atmospheres better than most other metals. As aluminum weathers, a grey oxide coating builds up at the surface and slows further corrosion. Anodizing is essentially an artificial thickening of this natural oxide. For more resistance to corrosion, standard aluminum sheets can be either painted or coated by dipping in a hot aluminum/zinc bath (Alclad aluminum sheets, available from most manufacturers of aluminum).

Like zinc, aluminum expands and contracts more than other roofing metals, and roofs should be designed to accommodate this movement. Aluminum also reacts galvanically with most other metals, especially in coastal or industrial atmospheres. Aluminum should be separated from all other metals except stainless steel, zinc, monel and lead. Also, water runoff from copper should be diverted away from aluminum.

Patterns—After you have chosen a metal to suit your fancy and your budget, the next step is to select a pattern. The crimped or folded pattern of a metal roof gives the metal rigidity, provides a system for joining adjacent sheets, and directs water off the roof. Besides those practical functions, the pattern also determines the style. Traditional roofing patterns form a series of strong lines from ridge to eave, giving a feel of rhythmic repetition as the pattern steps across the roof.

Metal roofing patterns are formed in one of two ways: brake-pressing or roll-forming. In the first method, panels are fabricated in a

Photos this page: J. Azevedo; Illustrations: Barbara Smolover sheel-metal shop with a press brake (FHB #9, p. 26), a massive but simple tool for making long, straight bends in a metal sheet. Metal in rolls or sheets is sheared to size and then bent into units called pans. The length of the pan is limited by the size of the brake, usually 10 ft., and the pan width is governed by economy and fastening requirements. Narrow pans, with more seams per square foot of roof, cost more. Wide pans (greater than 20 in.) leave too much room between seam fasteners. The wide, flat areas of metal will buckle visibly ("oil-can") with changes in temperature.

Pans are formed into one of two typical patterns: standing seam or batten seam (drawing, top right). Both of these interlocking patterns feature concealed clip fasteners, with no exposed nails or screws. The standing seam has a clean, sharp silhouette, while the wider profile of the batten seam looks good on large roof expanses. The batten seam requires about 20% more metal for the same coverage.

Today, most metal roofing is not formed on a brake in a local sheet-metal shop but rather in an industrial rolling mill. Rolling mills turn out roofing panels much faster than a brake, forming enough for a roof in minutes. Most manufacturers will custom-cut their roofing panels to run the entire length from ridge to eave, eliminating the builder's problem of sealing end laps. The practical maximum length of these panels is limited only by hauling restrictions (typically about 36 ft., although longer panels can be shipped with special permits).

Some professional roofers now roll their own panels with portable roll-forming machines. These scaled-down rolling mills, small enough to be towed behind a pickup, can be brought on site to form standing-seam roofing panels of any length (photo facing page, top). Once in place, the panels are crimped together either by hand or by a seam-forming machine that crawls along like a little robot (photo facing page, bottom).

With roll-formed panels, the pattern profile can be a series of straight lines (like those made with a brake), curves (corrugated), or any combination of the two (drawing, center right). The major differences between the various patterns are appearance and the way the panels are fastened.

Most roll-formed panels are nailed or screwed through the face of the panel to the roof deck. A covered neoprene washer seals the hole. Nailing is usually recommended only for outbuildings where leaks will not matter. Movement of a metal roof will eventually pop the nails, and water will seep down through the nail holes. Screws are better. The screw turns up a tiny lip, in effect a dam, around each hole. And of course, screws bite into the wood and hold longer.

Contrary to popular opinion, most metal roofing panels should be fastened in the troughs rather than through the crowns (third drawing from top). The long exposed stem of a screw in a crown means that the shank can work back and forth like a pendulum, which will loosen or break it. Screws in the troughs

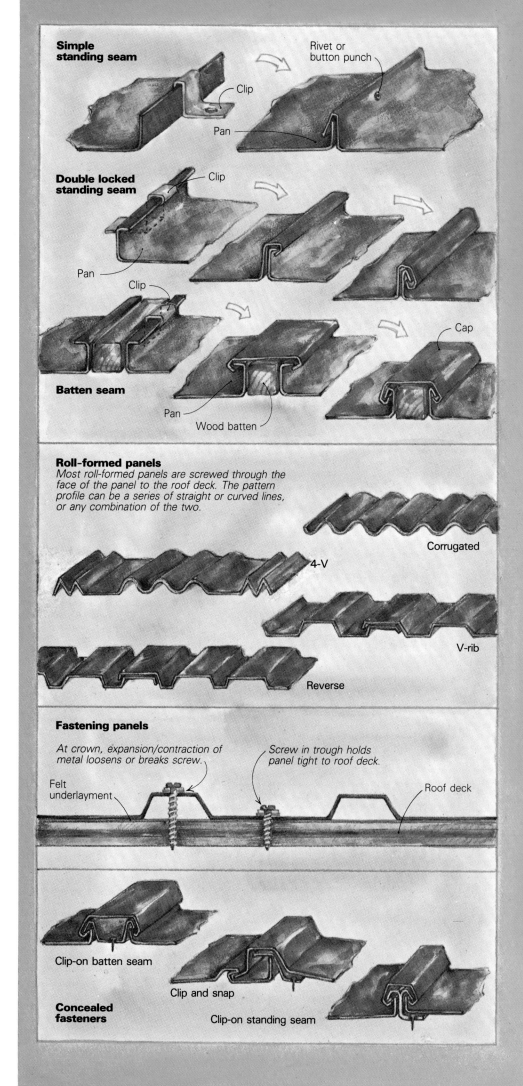

Corrosion

The word corrosion comes from the Latin *corrodere*, "to gnaw to pieces." Anyone who has seen a rusting steel roof will understand the accuracy of that description.

Metal can be destroyed either by chemical or by electrochemical corrosion. During chemical corrosion, metals react with oxygen, water and pollutants in the air, forming oxides at the surface. You can see these oxides as red rust on steel, white oxide on aluminum, white carbonates on galvanized steel and zinc, and green sulfates on copper. After initial weathering, this coat of oxides slows down further corrosion.

Just as metal auto bodies disintegrate under the attack of winter road salt, metal roofing succumbs more rapidly to coastal salt air or industrial fumes. For example, on the West Coast, many mountain cabins have plain galvanized roofs that show little sign of weathering after 30 to 40 years in the pure dry air, while 100 miles away on the seacoast, a five-year life would be considered quite good. Acids and salts in some woods also accelerate chemical corrosion, especially in moist environments. Corroded copper flashing along a line of cedar shingles is an example.

Electrochemical, or "galvanic," corrosion results when two dissimilar metals are put next to each other. In the presence of water and salt, reaction between the adjacent metals will produce a tiny electrical current, which dissolves the more reactive metal (the anode) without harming the more inert metal (the cathode). Electrochemical corrosion is faster than ordinary chemical corrosion because protective oxides never get a chance to build up at the surface. Instead, they are carried away with the current. Coastal or industrial atmospheres accelerate electrochemical corrosion.

Aluminum and zinc (including galvanized) are most susceptible to galvanic corrosion and should be separated from other metals except for stainless steel and lead. Wood or insulation that has been treated with copper or mercury preservatives will also react with aluminum and zinc. Other common preservatives, such as pentachlorophenol, creosote and zinc naphthenate, will not corrode metal roofing.

Protection against corrosion—

Metals susceptible to corrosion are either alloyed for better resistance (stainless steel), physically covered with a less reactive metal (terne), or covered with a more reactive metal (galvanized). The galvanized plating protects the steel core in two ways: it's a physical boundary that shields the steel, and it's more reactive so it breaks down more quickly than the steel—a sacrificial layer of molecules.

A metal roof needs periodic maintenance to keep corrosion at bay. From time to time, wash off leaves and debris. For paint finishes, check the coating for cracking or peeling. Pay special attention to the joints. If you find a small blotch of rust, brush it bare, prime it with zinc chromate or iron oxide, and touch up the finish coat. Repaint the entire surface regularly, before the roof acquires major acne. Finally, if this does not sound contradictory, try to stay off the roof. —*J. A.*

can be drawn tight to the roof deck. Consequently, most roofing manufacturers recommend fastening in the troughs unless the panel has rounded troughs (corrugated) or troughs that are spaced too far apart to meet the fastening schedule.

Manufacturers have devised various ingenious clips, snaps and caps to hold the panels securely while concealing the fasteners from the weather (bottom drawing, previous page). Some of these new designs cost little more than the standard panels, but offer the hope of a longer roof life with less maintenance. The future of metal roofing points toward roll-formed panels with hidden fasteners.

The last class of metal roofing patterns is not a panel but rather small stamped plates. The pressed geometrical shapes, based on Victorian scallop and fish-scale patterns and stamped from sheets of Galvalume, genuinely recall the grace of the old zinc shingles. Other metal plates are stamped and colored to mimic the appearance of wood shingles or tile.

Coatings—Metal coatings extend the life of roofing panels and generally improve their appearance. Paint, by far the most common protective coating for metal, can be applied either at the factory before forming or by hand in the shop or on the roof. Hand painting works fine for terne metal, which has an affinity for paint, but not for galvanized steel. Hand-painted galvanized roofing commonly flakes and peels after only a few years' exposure. The problem is preparing the galvanized surface for painting. Galvanized steel picks up a film of oil during manufacture and forming. This oily residue not only coats the surface, but also lodges within the pores of the galvanized plating. Unless the oil is removed, paint will not adhere to the metal.

A problem that affects imported steels is the chromate coating applied to them before they are shipped. This coating must be brushed off before painting, or the paint will not adhere. Alternatively, the metal can be allowed to weather a year before painting. Before you decide to try that, though, consider the difficulty of coating the metal hidden under the panel overlaps after the roof is already in place.

If you still want to hand-paint a galvanized panel, here is a preparation treatment that will give you the best chance for success. First, scrub the sheets front and back with a stiff brush and a strong, hot solution of trisodium phosphate followed by a rinse and a wash with dilute acetic acid (vinegar). Dry the metal and prime it with zinc chromate or red iron oxide. Finally, paint on the finish coat.

Buying prepainted roofing panels will save you a lot of work and may not cost any more. Also, the factory will do a better job because it coats the metal under ideal conditions. In a factory, long coiled strips of unpainted metal first get a vigorous alkaline scrub, then a phosphate pretreatment called a conversion coat, which is followed by an acid rinse. The prepared strip then passes through a series of coating rooms and ovens where the metal is primed, painted and cured. Finally, the strip is re-coiled and sent to the rolling mill.

Paints are available in various grades. The least expensive is the basic polyester enamel, similar to the finish on your car. It generally comes with a ten-year guarantee on film integrity and a five-year fade standard, which allows a slight change in color over five years. The silicone-modified polyesters, which cost about 20% more than the basic enamel, usually have a twenty-year guarantee on film integrity and a ten-year fade standard.

Fluorocarbon coatings, the next step up in quality, last longer and remain more flexible than polyesters. Kynar 500 (Penwalt Corp., Plastics Dept., 3 Parkway, Philadelphia, Pa. 19102) was the original fluorocarbon resin finish, but other manufacturers offer similar fluoropolymers, notably Duranar by PPG Industries (One Gateway Center, Pittsburgh, Pa. 15222) based on modified Kynar and other resins. These coatings typically come with a 20-year guarantee on film integrity and fade.

At least two roofing manufacturers—Berridge Manufacturing Co. (1720 Maury St., Houston, Tex. 77026) and Architectural Engineering Products Co. (AEP, Box 81664, San Diego, Calif. 92138)—now offer panels with an exotic copper-like coating. Ground copper suspended in either acrylic or fluorocarbon resin gives this unusual coating the look of new or aged copper. AEP's coating weathers naturally like copper, starting with a new-penny brightness and aging to a dark patina. Copper-resin coatings cost about the same as the premium fluorocarbon coatings.

The most expensive and most durable coating is not a paint but an acrylic film called Korad (Polymer Extruded Products Inc., 297 Ferry St., Newark, N. J. 07105). This 3-mil plastic polymer is stretched like Saran wrap over primed galvanized steel and then heat-laminated to the metal surface. The thick film resists ultraviolet weathering and stays flexible, without chipping or peeling.

How much does a metal roof cost?—In the metal-roofing market, an "average" price is an illusion. You can find a metal roof to suit essentially any budget. A good painted galvanized roof in a standard V-rib pattern will cost about the same or a little more than a wood-shingle roof or premium composition roof. Fancier styles and patterns with concealed fasteners might cost up to twice as much as a shingle roof. On a custom bending job, shop labor alone can cost more than an entire painted galvanized roof, and the whole custom job with a premium metal might run four to five times the cost of a shingle roof.

The important thing to remember about metal roofing is that front-end costs are not as telling as life-cycle costs. A properly installed and maintained metal roof can last the life of the house. Consequently, the average yearly cost of a metal roof is less than for any other type of roof covering. □

Technical-writing consultant J. Azevedo lives in Corvallis, Ore.

The Classic "Tin" Lid

Hand-forming a standing-seam metal roof

by Matt Holmstrom

Roof pans are laid from ridge to eave. The roof is created with seaming iron and hammer, 'turning' the edges of the pan into a standing seam.

With a few peculiar hand tools and a knowledge of some old-time techniques, an enterprising remodeler/restorer can produce one of the most attractive and longest-lasting residential roofs around—the standing-seam metal roof. While its clean lines and unmistakable hand-formed look are still admired by old house lovers, "tin roofing" has generally fallen into disuse, as roof after roof is replaced with composition shingles. This is unfortunate, because metal roofing is indispensable for solving some of the common problems of old-house roof work. For example, hand-worked metal roofing is essential when matching existing architecture with new additions. It's great for covering cornice returns, bay window roofs and other locations you don't want to roof again for many years. A standing-seam roof is perfect for roofs too flat for shingles, yet too steep for built-up roofing. Here in western Virginia a tin roof has been the roof of choice for many years.

A standing-seam metal roof is really quite simple. The idea is to fold sheets of metal together to create a weathertight skin, without penetrating the skin by fasteners. The metal is cut to length and bent into a "pan" having an angular, U-shaped profile. Unlike rows of shingles, pans run vertically from peak to eave. As each pan is laid on the roof deck, it is fastened with "nail cleats," cut and formed on site from the same metal as the pans. As the pans are laid edge to edge and fastened in place, the edges are double rolled, or "locked" together by hand to create a 1-in. high standing seam (for a description of tin-roof tools, see the sidebar on the next page). No sealants other than solder are used, except at flashings, where we sometimes use silicone caulking.

Choosing materials—A handworked standing-seam roof can be fabricated from either sheets or rolls of galvanized steel, terne (steel coated with a lead-tin alloy), copper or terne-coated stainless steel. For workability, terne and copper are best; galvanized steel is stiffer and more difficult to solder, and terne-coated stainless is stiffer yet (for a comparison of the properties of the different roofing metals, see the article on pp. 88-92). Copper and terne-coated stainless are the premium materials in terms of longevity—and cost. Terne (the vernacular term is "tin") is long lasting if painted every few years. Galvanized steel can be left bare for a few years, but then it too must be painted. Whichever metal you select, use only compatible materials for nails and flashing.

My choice for a roof, at least in this area, is the terne roof, because terne is readily available, very workable and easily painted (compared to galvanized roofing, anyway). It comes in coating weights of 20 lb. and 40 lb. (the weights refer to the amount of terne coating per 100 sq. ft. of metal). Twenty-pound terne is a flashing weight; forty-pound should be used for the roofing itself. The metal also comes in two thicknesses: 30-ga. steel for general use, and 28-ga. steel for heavier use. It's made by Follansbee Steel Corp. (State St., Follansbee, W. Va. 26037).

Terne comes in 50-ft. long rolls that are either 20 in. wide or 24 in. wide. The 20-in. width is best for residential roofing because it makes for a stiffer pan that is less likely to rattle in high winds. If you want to work with the wider stock, though, use 28-ga. steel for your pans to increase stiffness. All of the material comes with a brownish factory-applied shop coating, which keeps the metal from rusting while in storage.

Underlayment requirements—Terne roofing should be applied over a skip-sheathed roof deck. Condensation frequently occurs on the underside of a metal roof, and solid sheathing traps this moisture against the metal, causing it to rust from the backside. When decking a new roof, I use #3 1x6 or 1x8 pine boards, spaced about 1½ in. apart over rafters 16 in. o. c. I use the edge of a 2x block to gauge the space between courses. If you don't have enough support under the metal, it will be too weak to walk on and too bouncy to attach cleats to.

Make sure the attic is adequately vented to help control condensation and prevent moisture damage. Soffit and gable vents are best because commercial ridge vents and power fans don't mix very well with seamed metal roofing (they are usually designed for face nailing in place).

As another guard against trapping moisture under the roof, use red rosin paper instead of roofing felt beneath the roofing (photo below). This moisture-permeable material acts as a slip

sheet to help the roof pans expand and contract freely. Be careful not to step on rosin paper, though—it makes a good slip sheet for roofers, too. I get around this, particularly on steep slopes, by running the rosin paper from ridge to eave one course at a time as I run the pans. This makes it easy to work over, and the exposed skip sheathing offers plenty of toehold.

Estimating and getting started—If I'm using 20-in. wide terne rolls for a roof, my pans work out to be 17¼ in. wide (20 in. − 1¼ in. for one edge bend and 1½ in. for the other). Theoretically, a 50-ft. roll should cover almost 75 sq. ft. of roof area, but I allow about 12% for waste and for cutting cleats and flashings, which gives me about 66 sq. ft. of coverage for each roll. For edge cleats you'll need to cut 6-in. wide strips of metal from the long dimension of sheets of 24-ga. galvanized steel. This will give you 48 lineal feet of cleat from every 3-ft. by 8-ft. sheet of steel (it's also available in 4x8 sheets).

On old work I start by ripping off the old tin, but if I'm concerned about having too much open roof at one time, I'll remove the old stuff in sections, replacing it as I go. It's important to inspect the existing deck—there's no telling what you might find under an old roof—and repair any rot or damage in the rafters and sheathing. Set all nails in both new and old decking because a protruding nailhead can rub its way through expanding and contracting tin. Use a disc sander to smooth any other high spots.

Once the deck has been prepared, the edge cleats are the first metal to go on the roof, forming the first line of defense against wind damage. Cleats run along the entire perimeter of the roof, and will later be bent over the edge of the roof to form a drip edge (see E and G in the drawing on p. 96). When I nail up the cleats (every 8 in. with roofing nails or hot-dipped galvanized ring-shank nails), they should extend be-

(see E and G in the drawing on p. 96)

A few peculiar tools

As with most other crafts, having the proper tools on hand is half the battle. Unfortunately, most hand tinning tools won't be available at your local hardware or building-supply store, and clerks probably won't even know that such tools exist. Sheet-metal supply houses might have them, but at a handsome price. If tin roofs were once popular in your area, it stands to reason that tinning tools were too, and some real bargains may turn up at auctions. The cheapest route may be to

Turning a standing seam
The two outside edges of the seaming iron are of different heights, and alternately provide the anvil against which the hammer delivers blow after blow to roll the seams over. To use the iron, hold its taller side against the first pan. Flatten the second pan against the top of the iron with a mallet (step 3, below), then rotate the iron to the other side of the seam and pound the edge tightly over the edge of the first pan (4). This process is then repeated with the short edge of the iron to double roll the seam. The finished seam is a 1-in. high, double-rolled (or locked) standing seam (5).

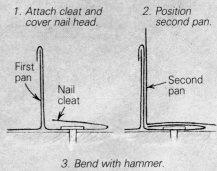

1. Attach cleat and cover nail head.

First pan

Nail cleat

2. Position second pan.

Second pan

3. Bend with hammer.

Iron

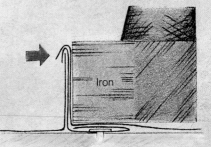

4. Flatten with hammer.

Iron

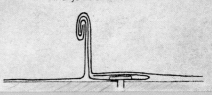

5. Repeat steps 3 and 4 with lower edge of the iron. Final joint should look like this.

Drawings: Chuck Lockhart

Tools of the metal-roofing trade. *Holmstrom uses tongs (left) to bend the edges of pans before forming a seam. Here, he will flat-seam roof pans into the pan that forms the hidden gutter. Note that the standing seams have been flattened near the work. This allows the pans to be bent without distortion. Next, Holmstrom 'turns' the seam with a seaming iron (above center) and hammer. The metal is folded in upon itself twice and pounded flat. This seam will later be soldered to increase its water resistance. A similar process is followed along the ridge, only instead of being pounded flat, the joint is left as a standing seam. Standing seams are generally formed with hammer and seaming iron, but a pair of foot seamers (above right) speeds the work. It's an antique tool, but thoroughly effective—the jaws of the seamers can bend and crimp the metal with foot pressure applied to the pedal.*

make the tools yourself—given the simple design and function of many tinning tools, this isn't as outrageous as it sounds. A friend of mine, for example, made a set of tongs for only $8; in contrast, a new set costs about $140 from a sheet-metal supplier.

The most basic tinning tool set consists of a set of tongs, a seaming iron and a hammer. Tongs (photo above left) are used to bend the sheet metal into roof pans. The jaws of the tongs have threaded holes in them for a pair of stop screws that determine the depth of the bend—you set the location of the screws depending on the depth of bend you want to make. Tongs will be among the most expensive hand tools in your basic tinning set, but you will use them a lot.

Valley tongs are similar to standard tongs, but are deep throated so the jaws can reach the center of a pan to make a bend. I don't have a pair of these because I've found something faster and easier: a portable sheet-metal brake like the ones used by aluminum-siding crews. With my 10½-ft. long brake, I can bend valley pans, flashings, nail cleats and shorter roof pans. Brakes are fairly expensive new ($400 to $900), but they are readily available used.

The seaming iron (middle photo, above) functions as sort of a mini-anvil, and has two machined steel edges and a metal handle. One edge is 1¼ in. high and is used for turning the first edge of a seam; the other edge is 1 in. high for rolling the seam a second time to lock it. To use the seaming iron, you simply set it along the seams between pans and hammer the metal against alternate sides to create the standing seam (drawing, facing page). You

can expect to pay about $40 for a new seaming iron.

The standard tinner's hammer is made entirely from hickory or ash. But it doesn't take long to wear down the face of a wood hammer, and if you don't keep it trimmed up to a 90° face, you'll have problems pounding tight seams without damaging the flat surfaces of your pans. For superior durability and control, I far prefer a Stanley Compo-Cast 2-lb. deadblow hammer (The Stanley Works, 195 Lake St., New Britain, Conn. 06050). It's a urethane hammer with steel shot in the head to reduce rebound, and will leave fewer hammer marks on your metal.

Good aviation snips—right-cut, left-cut, and offset—are indispensable because you'll have to cut a lot of metal. Don't bother with cheap snips or you'll be throwing them away out of frustration before the end of the job. I've found that Bergman makes a pretty good one (Bluebird snips, by Bergman Tool Mfg. Co., 1573 Niagara St., Buffalo, N. Y. 14213).

You'll need plenty of locking pliers—use the flat-faced kind made for working sheet metal (photo p. 93). Three or four are a start, but a dozen or two aren't too many for serious tin work. You'll need them for clamping pans together before and during seaming, for battening down sheet plastic and tarps to protect unfinished work and for all sorts of other uses. You can also use locking pliers for crimping and small bending tasks, but a pair of hand-seamers, available at sheet-metal or siding supply houses, works better. You can also get a hand-seamer (and a pretty good pair of snips, too) from Malco Products, Inc. (Annandale, Minn. 55302). Offset hand-

seamers are useful for working in those hard-to-get-at places.

You may not need a soldering iron for small roof jobs, but you probably will on the large ones. Commercial roofers sometimes use irons heated in charcoal braziers, but this seems primitive to me. I prefer electrically heated irons with at least 175 watts of power and a wide-bit tip—a standard soldering gun or soldering iron won't do the trick. You may find these high-wattage soldering tools at a sheet-metal or stained-glass supply outlet.

You might want to invest in a set of foot seamers—they're nice to have but not essential (photo above right). Foot seamers use both leverage and foot power to turn a seam and flatten it at the same time, and they'll eliminate hours of hammer and seaming-iron work on an average roof. They're delightful to look at and simple to use, but very hard to find. A friend of mine found an old set at an auction and paid $25 for it, but I've been quoted prices as high as $900 for a set in good working order.

I've found power shears indispensable, though only because I do a lot of metal roof work. Shears cut the heavier metal needed for edge cleats and without them, you'll need to have a sheet-metal shop rip your edge cleats. My Kett K-200 (Kett Tool Co., 5055 Madison Rd., Cincinnati, Ohio 45227) excels at cutting into old tin roofing—ripping off an old roof seam by seam can be brutally slow.

Finally, I keep some ladder hooks handy for any steeply pitched roof. These steel brackets attach to the top of any standard aluminum ladder and transform it into a very safe and lightweight roof ladder. —*M. H.*

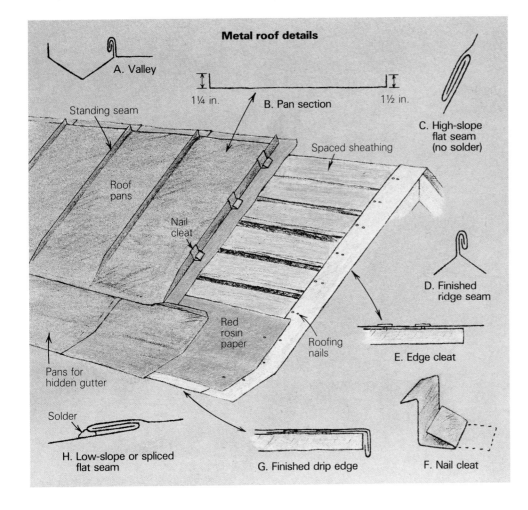

Metal roof details

A. Valley

Standing seam

1¼ in.   B. Pan section   1½ in.

C. High-slope flat seam (no solder)

Roof pans

Spaced sheathing

Nail cleat

D. Finished ridge seam

Red rosin paper

Roofing nails

E. Edge cleat

Pans for hidden gutter

Solder

F. Nail cleat

H. Low-slope or spliced flat seam

G. Finished drip edge

---

yond the edge of the roof by about 1¼ in. at the rake boards and about 2 in. at the eaves. The joints need only to be butted, not lapped.

Forming pans and nail cleats—On a simple gable roof, the pan length is determined by measuring the distance from the bottom of the eave edge cleat to the ridge, and adding 2 in. (1¼ in. for the ridge seam and ¾ in. to overhang and crimp to the edge cleat). I usually measure this length (adding a 6-in. allowance for error) while the old roof is still on, and bend up the pans beforehand. Then I can just trim the excess once the pans are up. But on a hip roof this isn't possible. Instead, snap a chalkline eave to ridge every 17¼ in. across the deck (this approximates the pan width), and running into the hip. Measure up to the "highest" point of each pan to find the rough dimension for cutting it.

Roll out the metal on a firm, flat surface and cut it to the rough pan length with snips or shears. With the tongs set at 1¼ in., grasp one edge of the metal in the jaws and bend it up about 45°. Continue along the entire length of the strip, then go over the edge again to get the entire bend up to a full 90° (if you're using a sheet-metal brake, you can do this in one operation). Reset the stop screw on the tongs to 1½ in. and repeat the process on the opposite edge (B in the drawing above). You'll probably find it helpful to put some weight in the center of the pan to keep it from sliding around as you bend it. The first pan to go up on a gabled roof is at the gable end, and it isn't bent in the U-shape that characterizes those to follow. Instead, one side is bent down ½ in. to lock onto

the edge cleat, while the other side gets bent up 1¼ in. for seaming to the subsequent pan.

To make the nail cleats, bend strips of roofing metal (F), then use the hand snips to cut 2-in. wide nail cleats from the strips. The sheet-metal brake works best here, but hand seamers or even locking pliers can be used. An average-sized roof may take a thousand or more cleats, so a tinner can always pass some time on a rainy day by filling up the cleat bucket.

Running a roof—Set the first pan over the rake edge cleat, with the upper end at least 1¼ in. past the ridge and the lower end overhanging the eave cleat by about ¾ in. Reach over with the hand seamers and firmly crimp the pan to the cleat down its entire length. This will later be bent downward to form a drip edge. Cut a length of rosin paper and tuck it under the pan, overlapping some of the edge cleat with the paper. Slip nail cleats over the outboard edge of the pan about every 12 in., crimp them tightly to the pan, then nail the flanges to the roof deck with a pair of 1-in. long roofing nails. Bend the back edge of the cleat back over the nail heads to protect the underside of pans and discourage nails from pulling out. At the bottom of the pan, don't bother with nail cleats—nailing through the 24-ga. edge cleats can be brutal, and the final bottom crimp will secure the pan here anyway.

On the second pan, and all subsequent pans until you reach the opposite end of the roof, you'll get lots of practice in hammer work. Each pan is placed next to the previous one, with its 1½-in. edge alongside the neighboring 1¼-in. edge, and with the same overhang at ridge and

eave. I clamp them together with locking pliers in several places, and then fasten nail cleats.

Use the seaming iron and hammer to turn the seam once along the entire length of each pan (photo p. 93). To cover the open roof as fast as possible, I double-roll the seam in only a couple of places to lock it in place. When I've laid all the pans on one side of the roof, I'll come back to put the second turn on all seams. After every two pans, I put down another length of rosin paper, lapping it only slightly over the previous length; a few roofing nails hold it in place.

"Losing" a seam is frustrating. While seaming, the top fold in the 1½-in. edge can get flattened without the top of the 1¼-in. edge inside it. This usually results from sloppy bending techniques (like an undersized 1¼-in. edge or a very uneven roof deck), but will occasionally happen even to the best roofer. If it happens to you, just pry up the flattened metal with a hammer claw or screwdriver, then carefully reseam it.

If the roof deck is out of square, the ends of the pans will "sawtooth" as you work your way across the roof. To minimize this, I carefully align the pans with the roof edges when I start a side, but after that I don't pay much attention to it. Later on, when the bottom edge of the pan is crimped around the edge cleat, the edge of the roof will look quite straight. The important thing is to make sure you have at least 1¼ in. of metal for the ridge seam. The ridge will be a compound seam, which is difficult anyway—one that's too short is that much tougher. If the roof deck is severely out of square, cut the pans longer to compensate and just trim the excess down to 1¼ in. once the pan is in place. At the end of the roof, another edge pan locks over the edge cleat. Running pans for a hip roof works pretty much the same way, though each hip pan must be trimmed to the angle of the hip.

Ridge details—Once all the seams on one side of the roof have been double-rolled, I flatten about 6 in. of the seam at the top and bottom of each pan by just beating it down with the hammer, in preparation for the compound seams to come. On hip roofs, flatten the seams in a downslope direction to prevent flowing water from pocketing under them.

Once the pan seams have been flattened, I can begin to form the ridge seam. With the tongs, I grasp the extra 1¼ in. of pan that extends past the ridge and bend it backwards, so that it's in the same plane as the opposing side of the roof. Then, after running all the pans on the other side of the roof, this lip is stood up with the tongs and double rolled with the opposing side (D). The second roll on one of these compound seams is quite stiff, especially where pan seams come together, so I'm careful to lay out the second side of the roof so as not to allow pan seams to coincide with those on the previous side. On a roof framed with an irregular hip, some seams will coincide no matter what you do, and you'll really have to work to turn them down.

Splicing pans—What's the longest pan you can comfortably work? I've seen and even installed pans up to 24 ft. long, but at this length, expansion can be a problem and handling can be a nightmare. Generally, I consider anything over 14 ft. to be too unwieldy to handle and go with a two-piece pan.

Roof pans that must be spliced are cross-seamed together. A ¾-in. "bend up" is put on the upper edge of the lower pan, and a ¾-in. "bend down" is put on the lower edge of the upper pan (C in the drawing on the facing page). These bends are across the flat bottoms of the pans only—the edges are left alone to overlap each other. The upper pan edges must sit inside those of the lower pan, and sometimes I have to bend the upper pan 1/16 in. narrower for this to work out. Once they're lodged together, the bends are flattened with a hammer and the pan can be nail-cleated and edge-rolled just as one-piece pans are.

A cross seam shouldn't fall over a gap in the skip sheathing, otherwise it can't be flattened without distortion. Cross seams in adjacent pans should be staggered. The face of the cross seam should be soldered, though this may not be necessary for steep roofs (I usually solder cross seams when the roof pitch is less than 8-in-12). On slopes of less than 6-in-12, a special cross seam is used that features a soldered cleat and a 4-in. minimum overlap.

You're probably wondering about the roofs you've seen with innumerable cross seams. The time it took to solder every seam was considerable, I'm sure, but the reason for this was more the availability of materials than the patience of the roofer. Metal used to come in short sheets because it was hand-dipped to get the terne coating, not continuously coated, as it is now.

Running valleys and compound seams— Valleys are installed as separate pans before any adjoining roof pans go on. Each one starts as a 20-in. wide or 24-in. wide sheet of 40-lb., 30-ga. metal (use 28-ga. on low slopes). The metal is first bent lengthwise down the middle to the approximate angle of the valley, then each edge is further bent inward 1¼ in. at a 90° angle (A in the drawing). I make the first bend either with the deep-throated valley tongs or on the sheet-metal brake, and the second two with regular tongs or the brake. The pan is then laid into the valley atop red rosin paper, and temporarily fastened to the perimeter edge cleat with locking pliers. The lower edge of each valley should be trimmed to overhang the edge cleats by ¾ in. Nail-cleat each side of the valley on about 12-in. centers, then flatten the sides inward. Pans coming into the valley are trimmed to a length ¼ in. past the edge of these flattened bends, then they're cleated and seamed to each other. When completed, the standing seams are flattened over the valley. Then with tongs or a hand seamer (on a steep valley the tongs may not fit), the flattened valley edge and the pan edges are stood up and double rolled toward the valley. Usually the doubled valley seam is flattened, and on a lower slope it's also soldered. Splices in valley pans should be soldered.

The trickiest spots on a roof are where compound seams meet, as where a hip runs into a ridge or two valleys meet at a ridge. Usually the downslope seam is completed and flattened out

where it is to become part of the upper compound seam, and the upper seam is completed afterward. An exception is at the eave edge cleat. Hip or valley seams coming down to the edge of the roof are completed and flattened at the eaves. The entire roof edge is bent around the edge cleat and crimped with hand seamers, or with hammer and seaming iron. (Use locking pliers to clamp the bottom of the pan as you work along it.) Then the whole edge is bent down. On the rakes, I bend the overhang flush with the fascia board. On eaves, I angle the edge away from the fascia some, for good drainage into a gutter. Since folds are made through both roof metal and edge cleat, you have to be forceful with the hand seamers and the hammer.

Changing slopes—Standing-seam metal roofing works well on a roof with a slope change. But if you try to bend a roof pan across the flat of the pan, the bent edges will fold up on themselves and the bottom of the pan will crease. The trick is to devise a tool that controls this folding—I use part of a broom handle for the job (drawing at right).

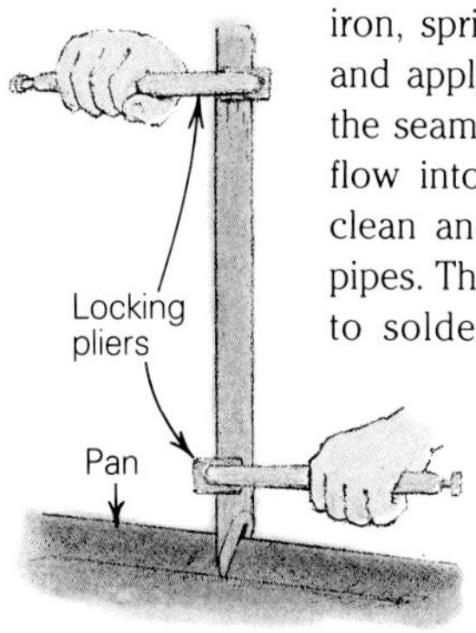

To use the tool, place the slot over the standing bend of a pan where the slope should change, and turn it until the pan flexes toward the desired angle. Repeat this on the other side of the pan. Since angled pans aren't able to expand freely, it's a good idea to splice them on the upper slope.

A standing-seam roof becomes a risky proposition at pitches less than about 1½-in-12. Although I have read recommendations limiting standing-seam work to minimum slopes of 3-in-12, this seems too conservative to me. I once tore off a 1-in-12 standing-seam roof on a small porch that was underlaid with 1910 and 1911 newspapers. Naturally, I replaced it with standing seam, figuring that 71 years of leak-free service proved a point. But it's a judgment call and on a large roof, or one with other roofs draining onto it, you should use a flat seam throughout.

A flat-seam application is essentially a standing seam turned once, flattened, and soldered (H). The pans are bent with a 1¼-in. edge and a 1-in. edge, so only the low side of the seaming iron is needed to roll them. All low slopes are done in a flat locked seam that is soldered, but flatwork can be tricky because one popped solder seam means a bad leak. Whenever possible, I try to increase the roof pitch instead.

Soldering—Soldering can be difficult and very time-consuming, so first summon up all your patience, and then select the right flux. Flux is a material that helps metals fuse, and also prevents surface oxidation. I use a liquid flux called Stay Clean (J. W. Harris Co., 10930 Deerfield Rd., Cincinnati, Ohio 45242) for galvanized steel. Crystalline rosin flux is used for copper, new terne, and terne-coated stainless. I've had a tough time finding this material lately, but you might be able to get it from a sheet-metal supplier. Solder won't flow on weathered terne or

terne that has been stripped of paint (as in repair work), so I use a liquid flux called Rubyfluid Soldering Flux (The Ruby Chemical Co., 68-79 McDowell St., Columbus, Ohio 43215). In a pinch, the lead-tin coating on new terne will solder without any flux.

I use only 50/50 solid-core bar solder, never an acid or rosin-core solder. Wire solder can be used, but bar solder is more efficient for tinwork. Though 1-lb. bars are the most common, I prefer ¼-lb. bars because they seem to get just the right amount of solder on the seam without gumming up the tip of the soldering iron. An average roof might take about 2 lb. of solder.

Good soldering technique comes from common sense and practice. Heat the soldering iron, sprinkle or brush some flux on the seam, and apply the tip of the iron and the solder to the seam. If the seam is hot enough, solder will flow into it, though the process isn't quite as clean and easy as it is when sweating copper pipes. The flatter the seam is, the easier it will be to solder. Cross seams on steep slopes are tough because the solder wants to flow downhill rather than along the seam. To minimize this, use a dam of scrap metal to keep the solder from running down the pans.

Flashing—Flashing a metal roof is simpler than flashing a shingle roof because step flashing is eliminated. If a roof ends along a sidewall, for example, the edge of the last pan is simply bent 2 in. or so up the sidewall as a base flashing. To accommodate expansion, this lip isn't fastened to the sidewall in any way. Counterflashing, made from the same materials as the roof, is then slipped under the siding and lapped over the base flashing. You'll have to remove or pry up some siding in order to nail the counterflashing in place.

Where a roof slopes away from a wall, such as on a shed addition, there are two flashing options. One way is to cut the edges of the pan about 2 in. from the end, flatten them and then bend this "flap" to conform to the wall surface. The problem with this method is that it's hard to get solder into the small hole at the back edge of each standing seam. The second choice is to use a separate flashing pan. Stop the roof pans about 12 in. short of the wall, and seam them together. Put a 1¼-in. bend-up at the top edge, then bend a flashing pan to lock into it.

Chimneys and curbs are flashed by turning the pan up as base flashing, but tabs must be soldered in at corners to make the base continuous. Stepped counterflashing where a roof meets a chimney can be replaced with a single L-shaped piece of metal, tucked into an angled kerf cut in the masonry. Seal the kerf, and any nailheads through the counterflashing, with paintable silicone caulk.

For flashing vent stacks, I like to use prefabricated pipe collars with a neoprene gasket and a galvanized steel base. Bends should be made in the base so that it can be flat-seamed and soldered right into the roof pans. □

Matt Holmstrom is a remodeling contractor in Bedford County, Va.

Building and flashing a cricket are the same whether for a masonry chimney or for a wooden chimney structure like the one above, which houses a pair of woodstove chimney pipes. Other metals will work for the flashing, but copper is the easiest to solder.

Chimney Cricket

Soldered copper flashing for a rooftop watershed

by Scott McBride

The best place for a chimney to exit a roof is at the ridge. Water will run away from the chimney, not toward it, and flashing becomes a fairly simple matter. Builders in New England did this back in the days before sheet metal was readily available.

When you move a chimney downslope, however, runoff will strike it, often forcing water up under the flashing and down into the house along the bricks. Leaves and other debris can build up behind the chimney and decompose into a moisture-laden compost that rots roof boards and corrodes the flashing. The way to avoid this mess is to build a well-flashed cricket.

A cricket, or saddle, is a miniature gable roof covered with metal that sits between the back of the chimney and the main roof. The intersection of cricket and main roof produces a pair of valleys that divert water, snow and debris around the chimney. Recently I built a small cricket for a wooden chimney structure

that houses a pair of woodstove chimney pipes. Although the wooden chimney structure made attaching the cricket easier, constructing the cricket and flashing it were much the same as they would be for a masonry chimney.

Carpentry—The cricket I built was small enough to make with plywood—no rafters— but I'll speak in terms of common and valley rafters in order to relate the geometry to standard roof framing. Since the chimney structure was 2-ft. wide across the back, the run of the cricket common rafter was half of that (12 in.). The main roof of the house had a pitch of 9-in-12, so to make things easy I gave the same pitch to the cricket. The run of the cricket common rafter being 12 in., the rise was 9 in. and the length, or slope, an even 15 in. You can calculate this using the Pythagorean theorem, but it's easier just to measure the diagonal on a framing square.

Each half of the cricket roof surface (bottom drawing, facing page) is a right triangle with one leg 12-in. long (the length of the ridge) and the other leg 15 in. long (the length of the common). The hypotenuse connecting these two sides is the valley. To begin construction, I took a piece of ½-in. CDX plywood 12 in. by 15 in. in size and cut it diagonally, giving me both halves of the cricket at the same time.

To make the pieces fit properly, I beveled the ridge edges at 37°, which is the equivalent of a 9-in-12 common-rafter plumb cut. Then I beveled the edge that runs along the valley at 45°. This was a sharper bevel than was necessary, but the resulting undercut wasn't a problem.

I held the two pieces of plywood in place on the roof and traced the outline of the cricket onto the roof sheathing and the chimney structure. One-half inch below the lines on the chimney, I drew parallel lines marking the top of my nailers. To make these, I cut a

Flashing layout

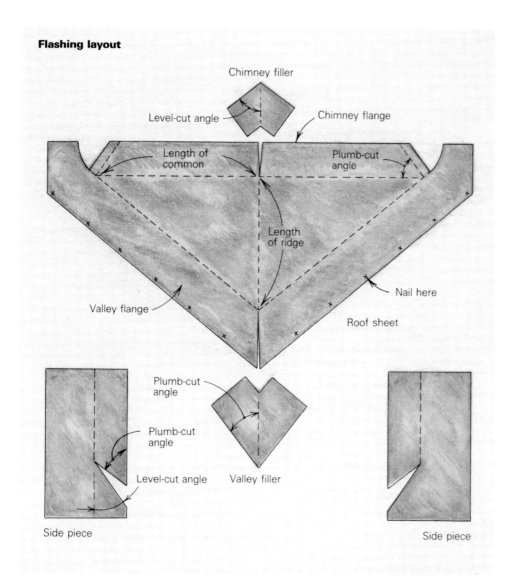

Chimney filler

Level-cut angle

Chimney flange

Length of common

Plumb-cut angle

Length of ridge

Nail here

Valley flange

Roof sheet

Plumb-cut angle

Plumb-cut angle

Level-cut angle

Valley filler

Side piece

Side piece

Framing and installation

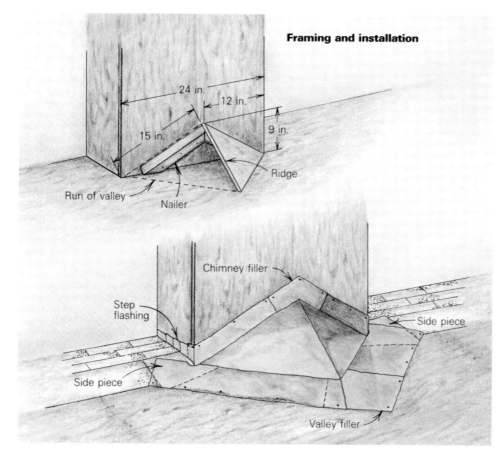

24 in.

12 in.

9 in.

15 in.

Ridge

Run of valley

Nailer

Chimney filler

Step flashing

Side piece

Side piece

Valley filler

pair of 2x4s with a 9-in-12 plumb cut at the top and a 9-in-12 level face cut at the bottom and nailed them in place. I didn't bother with an edge bevel at the bottom since the nailers wouldn't actually have to bear on the roof (drawing, below left). Had this been a masonry chimney, however, these pieces would have been cut as actual rafters. This would have required compound miters on their bottom ends, consisting of a 37° bevel in conjunction with the level face cut. They could then be installed an inch or two away from the brick for heat clearance.

Flashing—After nailing down the plywood, I laid out the copper flashing for the cricket. You can use other metals for this, such as aluminum or galvanized steel, but copper is the easiest to solder, and soldering is the best way to seal the joints. Copper comes in sheets and rolls, and is available in different weights and tempers. For this job, I bought 14-oz. hard copper from a local roofing supply. It comes in 3-ft. by 8-ft. sheets, and I paid $45 per sheet.

The flashing assembly consists of one large piece and four smaller ones (top drawing, left). For the large piece, I started with the same dimensions obtained for the plywood, then added 4 in. for a flange against the chimney structure and 8 in. for a flange against the roof. Vertical surfaces, such as the sides of the chimney structure, don't need as much protection as the roof surfaces, which is why I made the chimney flange so much smaller than the valley flange.

Two of the smaller pieces are fillers, covering gaps that open up at the chimney flanges and at the valley flanges when the large, main sheet is folded at the ridge line. The chimney filler was given a notch that was twice the angle formed at the top of each nailer. The valley filler was given a notch that was twice the angle between the valley and the nailer on the cricket's roof surface.

The other two pieces of flashing are side pieces that tie the valley flange, chimney flange and roof sheet together where they meet. The side pieces also turn the corner of the chimney structure and overlap the step flashing coming up along the chimney's side.

To make the side pieces, I started with a rectangular piece of copper and cut a slot in from one edge at 37° (the plumb cut angle). This allowed a flange to be turned up that would butt into the upturned chimney flange. I left tabs on the ends of the chimney flanges to create an overlap with the side pieces.

Most of the bending (top photo, next page) was done on a portable aluminum J-Brake (Van Mark Products Corp., 24145 Industrial Park Drive, Farmington Hills, Mich. 48024). But after making a series of bends, I could no longer slip the roof sheet into the brake and had to form the ridge crease over the edge of a 4x4 post. I also used hand seamers for some of the smaller bends. These are like pliers with broad flat jaws and are made specifically for working sheet metal.

To facilitate assembly and ensure accurate positioning of the pieces while soldering, I built a mock-up of the chimney structure. The mock-up allowed me to work in the shop and keep the heat of soldering away from the building itself—something that's always made me a bit nervous when soldering flashing in place.

Soldering copper—The secret of the tin-knocker's art lies not in the soldering itself, but in the fabrication of the parts. Close-fitting joints will suck in the molten solder just like joints in copper plumbing do, which makes sound, neat joints requiring minimal heat. Big gaps, on the other hand (anything over 1/16 in.), will require lots of solder and repeated heatings to seal them. Also, the excess heat causes the metal to warp, which drives the mating pieces even farther apart. The result is a thick, sloppy bead that may not be watertight. Once this happens, it's too late to go back and refit the copper, so the job has to be done right the first time.

After cutting out and bending the pieces, I cleaned the contact surfaces with steel wool (emory cloth works, too). New copper doesn't need much of this, but old tarnished copper requires mountains of elbow grease to get clean. Next, I carefully applied flux only to the areas where I wanted the solder to stick. Limiting the application of the flux in this way helps contain the molten solder once it starts to flow and enables me to peel hardened solder from those areas left unfluxed. I brushed on a liquid-type acid flux, as opposed to the paste flux preferred by plumbers. The liquid leaves less residue and is easier to wash off. I'm very careful not to get it on me—especially not in my eyes.

Before soldering each joint, I arranged the pieces so that the joint to be soldered was level. This reduced the tendency of the liquified solder to run away from the joint. I then clamped the pieces in place, keeping the clamps far enough away from the joints so they wouldn't act as heat sinks. Locking pliers, especially the deep-reach kind, are my favorite clamps for this. To keep the joints tight, I had a helper press down on either side of the flame with a pair of screwdrivers and move along with me as I soldered. This type of "moving clamp" is a great help.

The solder used by roofers is half lead and half tin. It comes in 1-lb. bars about 10-in. long. Traditionally, roofers melt it with a copper soldering iron heated in a charcoal-fired brazier. I've never had good results with this tool, however, so I use a propane torch. The drawback to the hardware-store-variety torch is that most are cheaply made—some of the ones I've used barely have outlasted the fuel canisters they were mounted on. An alternative is the heavy-duty air-acetylene torches commonly used by plumbers.

After adjusting the torch to burn with a sharp blue flame about 1½-in. long, I use a circular motion and play the tip of the flame over a broad area to warm up the copper a

bit. As I do this, I hold the bar of solder close to the flame so that it will also catch some of the heat. When the flux starts to sizzle, I put the tip of the flame down onto the work at the beginning of the joint. Then I press the tip of the solder bar against the work about ½ in. from the end of the flame and let the heat flow through the copper to melt the bar. I move the flame along the joint, followed by the solder bar (photo right). When you do this, you have to move at an even, moderate pace. If you move the flame too fast, the solder won't flow, but you can't linger either or you'll warp the metal with too much heat.

Textbooks on soldering will tell you that the heat should always be conducted through the work as I've just described, and that you should never melt the solder directly with the flame, dripping blobs of molten solder onto the joint. Well, that's fine if you've got a perfect joint lying flat on the bench.

But sometimes—especially if you have to work up on the roof, with your pieces at an angle, or if the pieces don't fit real tight—there won't be enough capillary action to keep the liquified solder from running out of the joint as fast as it runs in. The only practical way to deal with the situation is to pull the heat back a little so that the copper is hot—but not *that* hot. If you then drip molten solder onto a slightly cooled joint, it'll stick. It might not be as strong as a properly soldered joint, but at least it won't leak. And it's still better than a dollop of roofing tar.

After soldering, you have to clean off any remaining flux with water and a stiff brush in order to prevent corrosion.

Installation—Before cloaking the wooden cricket in its copper cape, I shingled up the sides of the chimney structure, inserting step flashing as I went. When I got up to the back

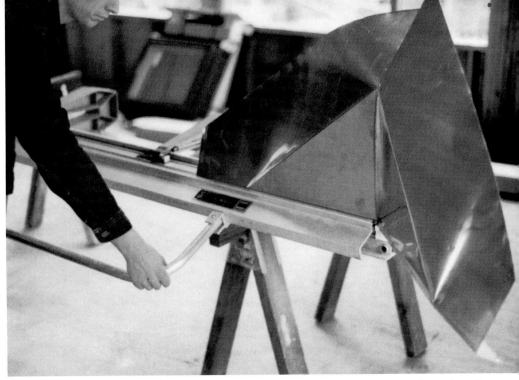

Most of the folds were made on the metal brake shown above. Eventually the configuration of bends prevented the roof sheet from slipping into the brake, so the final folds were handmade.

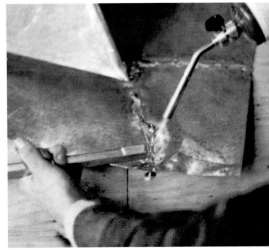

McBride solders the flashing in his shop, instead of on the roof. This way he can position the flashing so that the seams are level when he solders them, which keeps the molten solder from running out of the joint.

of the chimney, I slipped the sidepieces of the cricket flashing down over the uppermost shingle course and the top piece of step flashing. I anchored the cricket flashing with roofing nails sunk along the outside edges of the valley flange and chimney flange only (photo on p. 98). Then I continued with the roofing, bringing my shingles up to the centerline of the valley, but keeping all roofing nails off the flashing itself.

Since this was a wooden chimney structure, the vertical fins of the step flashing and the chimney flange were eventually covered by siding. Had this been a masonry chimney, a separate cap flashing, tuckpointed into the brickwork, would have been used as coverage. □

Scott McBride is a carpenter in Irvington, N. Y. and contributing editor to Fine Homebuilding. *Photos are by the author.*

Single-Ply Roofing
Tough, versatile membranes tackle the toughest roof problems

by Harwood Loomis

Since man first sought shelter from weather and carnivorous neighbors, he has preferred to have a roof over his head. Historically, the form of the roof has ranged from animal skins and thatched tree fronds to modern synthetic materials. The material most used in the U. S. for residential roofs these days, however, is probably the ubiquitous asphalt shingle. Other roofing materials include slate, clay, terra-cotta and concrete tiles, metal sheets, and wood shingles and shakes. All these materials have one thing in common: they are generally suitable only for use on roofs having a slope of 4-in-12 or more. At slopes below that, the designer and contractor have, until recently, had only one serious choice: the built-up roof, a multi-layer sandwich of roofing felt and either asphalt or coal-tar pitch (rolled roofing should only be considered for outbuildings and utility sheds).

Today, however, the conventional built-up roof is being eclipsed by a relatively new generation of sheet products called "single-ply membranes." Though associated mostly with huge commercial roof projects, single-ply membranes are now finding considerable use in residential construction.

The term single-ply membrane includes a vast number of different systems and materials, some of which are actually made up of more than one ply (photo next page). They're called single-ply because they are generally *installed* as one ply, rather than layered on like traditional roofing materials.

A single-ply roof is typically sold as a system of proprietary fasteners and adhesives which, along with the membrane itself, are applied by contractors licensed by the membrane manufacturer. It's possible for a builder to install a small job with materials purchased from a roofing contractor, but the membrane manufacturer typically won't extend any warranty on the materials in such cases (one exception is Resource Conservation Technology — see sidebar, p. 103).

Why single-ply—New single-ply membranes seem to come on the market daily, and there are easily more than 300 products to choose from. One of the attractive aspects of single-ply roofing in general is that it does away with the need for the smelly tanker full of molten asphalt that's needed for the installation of built-up roofing. And by getting rid of the ket-

Single-ply membranes are particularly useful in sealing flat roofs over living spaces. In the photo above, an EPDM membrane covers a sunroom addition to a large house.

tles, builders also get rid of the open flames used to heat them, making a safer job site. But there are other advantages as well.

Designers and builders of custom homes no longer need to have their imaginations limited by the roofing technology of decades past. Many single-ply products are elastic and allow roofs to take on complex, compound curves or to be nearly flat, yet still be watertight and durable. Some of the systems are well suited to double duty as roof terrace walking surfaces; others can serve as a waterproofing substrate beneath wood decks. Single-ply membranes are also effective when used to waterproof pocket decks and additions with flat roofs (photo above). Very probably, one of these products will be suitable for any roof configuration where shingles, slates or tiles can't do the job.

An alternative to built-up—To appreciate the popularity of single-ply roofing these days, it helps to know something about the evolution and use of built-up membranes, the traditional product for covering low-slope roofs. This is what some people refer to as a tar-and-

gravel roof, although it may not be made up of tar and it may or may not be surfaced with gravel. Typically it was composed of layers of roofing felt alternated with mopped-on layers of either melted asphalt or coal-tar pitch. This "old faithful" in its several variations has served builders well, but it has some drawbacks.

One problem is that the raw materials available today are not as good as they were in the past. The most common component of a bituminous roof is asphalt, which comes from various grades of crude oil. With prices rising as fast as they have lately, coupled with diminishing reserves, there is more competition to refine as much "product" as possible from the raw crude. Roofing asphalt comes out low on the totem pole in comparison to products such as gasoline, diesel fuel and heating oil. Many roofing experts believe that by the time all the other products have been drawn off, there isn't much left of what used to make roofing asphalt a satisfactory weatherproofing medium.

The best of the old tar-and-gravel roofs were made of coal-tar pitch. These roofs were excellent; I've inspected many that lasted 40 or

Though there are many kinds of single-ply membranes, the ones most commonly used for residential applications are shown above. From front to back: MBR (this one is styrene butadiene styrene), PVC, EPDM, reinforced PVC, EPDM, and CSPE.

50 years before beginning to leak. Unlike asphalt, coal tar becomes self-healing in warm weather—the warmth softens the coal tar enough to seal any cracks—so the roofs tended to repair themselves every summer. Unfortunately, the fumes from coal-tar pitch heated during installation have proven to be carcinogenic. This led the government to place strict limits on amounts and concentrations of fumes that could be given off during the installation of a coal-tar roof. That, in turn, resulted in changes to the formulation of the pitch itself, which diminished its durability. Coal tar isn't plentiful today, and most roofers and roofing consultants agree that what *is* available is not as good as the product of 40 years ago.

Another problem with a bituminous roof—perhaps "challenge" is a better word—is that residential designers today are creating roof shapes that are more irregular and complicated than they used to be, and they're coming up with all sorts of other low-slope roofs, including some with compound curves. The traditional bituminous roof just isn't up to covering shapes like these, for good reason. The strength of a bituminous roof is imparted by the layers of standard 15-lb. roofing felt that are rolled over the roof between moppings of melted bitumen (asphalt or coal-tar pitch.) The felts come in rolls 36 in. wide and roll out very nicely on a regular, flat roof deck. But they do not readily lend themselves to rolling out on curved or compound-curved surfaces.

Given the increasing problems with built-up roofing, it's no surprise that single-ply products are taking over the market for low-slope roofing. Here's a look at the major types of single-ply roofing membranes that have applications for the residential market.

Rubber membranes—One of the most popular materials for single-ply roofing is synthetic rubber. There are a number of synthetic-rubber products on the market, two of which seem to have established themselves as reliable, high-quality materials for roofing membranes: EPDM (ethylene propylene diene monomer) and CSPE (chlorosulphonated polyethylene). Hypalon, DuPont's tradename for CSPE, has become virtually synonymous with CSPE roofing. Both are marketed by major manufacturers and are available with a choice of several different installation systems to address various roof configurations.

EPDM comes in a black sheet that looks and feels like a tire inner tube (photo above). It's usually 45 mils thick (a mil is $\frac{1}{1000}$ in.) and weighs about .28 lb./sf. Sheets range from 10 ft. to 100 ft. wide. A sheet of EPDM is usually not reinforced with fibers, and that allows it to stretch somewhat to accommodate irregular roof shapes. EPDM is reasonably resistant to many chemicals but is susceptible to attack by organic fats (such as cooking oils), some solvents (aromatic hydrocarbons like gasoline and oil), and fire. A special formulation of the product, sometimes called FR sheet, is available for projects in which the membrane surface will be uncovered and where a fire-resistant roof is required.

EPDM membranes aren't available in colors, but they can be coated with liquid Hypalon rubber, which comes in a range of colors. Some manufacturers offer a white EPDM sheet, but they won't warrant it for as long as they would an otherwise identical black membrane. That's because the material that makes EPDM black—carbon black—also contributes to its durability; white membranes lack this element.

Installing EPDM—An EPDM roof may be installed in several ways. In a *ballasted* application, a large sheet of the membrane is laid over the roof deck and smoothed into place. A heavy fabric is placed over the rubber to protect it, and the assembly is then covered with ballast—washed, round stones or cast-concrete pavers—that keeps the membrane from lifting off the roof in high winds and protects it from fire. The ballast generally weighs 10 lb. to 15 lb. per square foot. The perimeter of the roof and any penetrations, such as chimneys or plumbing vents, are sealed with special adhesives and flashing material (an uncured and elastic form of the EPDM rubber itself). Seams in an EPDM roof are sealed with adhesive.

The advantages of this system are fast, simple installation and relatively low cost. The disadvantage is that the house must be strong enough to support the weight of the ballast material in addition to whatever loads have to be anticipated for wind, rain, snow and people. The ballast also makes it difficult to find and repair leaks, and it is often difficult to obtain round, unbroken stones in the size needed for ballast.

A second method for installing an EPDM membrane is *mechanical fastening*. There are a number of different systems for mechanical attachment, athough most are proprietary. The EPDM membrane is rolled out over the roof deck and fastened to the structure at regular intervals with either batten strips or some sort of metal or plastic fastener plate that is screwed to the roof deck.

The advantages of mechanical fastening systems are their relatively fast installation, moderate cost, light weight and the ease of finding leaks. A disadvantage is that wind action on the loose membrane may cause the fasteners to loosen and "back out" of the roof deck, causing localized fatigue of the membrane at the fasteners.

The third major method of installing EPDM membrane is called *fully adhered*. Rigid insulation board is first installed and firmly attached to the roof deck, usually by screws. The membrane is then rolled out and fastened to the insulation board with proprietary contact adhesives specially formulated for the purpose. These roofs are susceptible to wrinkles during installation, so a heavier membrane (60 mil) is used.

Like other installation systems, the fully adhered approach isn't perfect. Advantages include light weight and the ease of locating and repairing leaks. On the downside, the fully adhered system is the most expensive of the three methods and takes the longest to install. It leaves the membrane exposed to fire and physical damage and calls for more field seams than with other systems. This is because of the difficulty in working with wide sheets coated with adhesive—it's pretty tough to keep from getting wrapped up in your work. So manufacturers of fully adhered systems limit the sheet to a 10-ft. width. And if the roof will be a visible design element, high quality workmanship becomes very important; it's easy to wrinkle a fully adhered system while installing it.

The fourth method of installing an EPDM roof is called a *protected-membrane* system. Also called an IRMA system (Inverted Roof Membrane Assembly), it's basically a variation of the ballasted roof and is sometimes called an "upside-down" roof. The membrane goes over the roof deck, then Styrofoam insulation at least 1 in. thick goes over that. The Styrofoam is held in place with ballast; channels molded into the bottom of each Styrofoam sheet allow water to drain freely from the roof. The big advantage of this system is that it protects the membrane from thermal shock (stresses induced by daily and seasonal temperature swings). Thermal shock will speed the eventual deterioration of any roof membrane.

The CSPE membrane—DuPont's Hypalon is the best-known member of the CSPE family. Like EPDM, it is a synthetic rubber membrane, but Hypalon is reinforced with a polyester scrim, an open grid of fibers. The reinforcement increases the tensile strength of the membrane, but reduces its elasticity as well, making it less desirable than an unreinforced product for use on roofs with compound curves.

The standard white color of Hypalon (other colors are available) reduces rooftop heat gain,

but white is also difficult to keep clean in dirty, industrial sections of the country. Hypalon should not be selected purely for appearance reasons unless the homeowner is willing to expend some effort to keeping the roof surface clean (periodic scrubbing with household detergent should do it).

Hypalon roofs can be installed with any of the methods used to install an EPDM roof (ballasted, mechanically fastened, fully adhered and protected membrane). Rather than being bonded with adhesives, however, Hypalon seams are welded together. A hot air gun is used to soften the rubber, and the seams are rolled together to fuse the joint.

Hypalon is very resistant to a wide variety of chemicals, but it's susceptible to attack from oils and gasoline. Hypalon costs somewhat more than EPDM and requires more field seams (the widest Hypalon sheets are generally 5 ft. wide, while the *narrowest* sheets of EPDM are usually 10 ft. wide). It weighs about .29 lb./sf.

Other rubber membranes—There are a number of other synthetic rubber materials used as roofing membranes, although none seem to have achieved the widespread acceptance of EPDM and CSPE. Nonetheless, each has advantages under certain conditions. Among the other products are: CPE (chlorinated polyethylene), PIB (polyisobutylene), and EIP (ethylene interpolymer alloy).

A thermoplastic, CPE comes from the same base as CSPE, but it doesn't contain sulphur. It is extremely resistant to most chemicals. Field seams and repairs are heat-welded, and repairs don't require liquid reactivators to soften the original membrane. A CPE job requires that metal flashings be precoated with liquid CPE. Only one company sells this product right now, and they don't issue warranties on residential applications.

A PIB membrane comes as a 100-mil sheet

Most single-ply membranes can be applied without the need for heat. One type of modified bitumen membrane, however, employs heat to liquefy and adhere the underside of the membrane to the substrate.

with a fleece backing that can be set either into hot asphalt or a cold adhesive. It's compatible with asphalt, and relatively simple to install. The sheets have 2-in. wide self-adhering tape built into the edges; to seal a seam, a release paper is pulled away, the edges of the sheets are overlapped and the seam is pressed with a metal roller.

Like Hypalon, an EIP membrane is made from DuPont polymers. It's a thermoplastic consisting of large polymer molecules that, according to the manufacturer, lock in the plasticizer molecules to reduce plasticizer migration (more on this later). Field seams are made by heat welding; repairs can be made by heat welding without requiring the use of liquid reactivators. The system is available only for mechanical attachment.

Plastic membranes—With continued advances in the composition of roofing membranes, it's getting tough to differentiate between "rubber" and "plastic" membrane materials. While EPDM and CSPE are almost universally accepted as rubber, some of the other membrane materials listed above can be

argued to be plastics rather than synthetic rubbers. But the distinction between rubber and plastic, as far as I am concerned, is that a plastic membrane requires chemical additives called plasticizers to remain flexible. These plasticizers gradually outgas over time, and if plasticizer migration is excessive, the membrane may become brittle and crack.

One membrane that absolutely belongs in the plastic category is PVC (polyvinyl chloride). The earliest PVC roofs in this country were characterized by a loss of plasticizers and a resultant brittleness and shrinkage. They sometimes failed prematurely either because they cracked, or because they shrank and pulled away from the perimeter of the roof. Manufacturers of PVC membranes have devoted considerable time and effort to optimizing the amount and composition of the plasticizers they use, and claim to have corrected any problems. While this may be true, one major supplier of PVC membranes limits the warranty to 10 years and the other limits it to 15. By contrast, most EPDM suppliers will back warranties of up to 20 years for commercial installations.

A PVC roof can be installed using the same techniques as those used with EPDM roofing (although one major supplier does not include a fully adhered system in its line). Seams are either fused with solvents or welded with hot air. Like EPDM and CSPE, PVC roofs are lightweight (except for the ballasted systems) and are easy to install. Cost is generally in the same range as EPDM.

Another problem with PVC membranes, however, is that they are less resistant to chemicals than EPDM or CSPE. In addition to being susceptible to attack from oils and animal fats, PVC can also be damaged if exposed to bitumens used in conventional roofing. This means that if PVC is used to reroof an area already covered with asphalt shingles or con-

Sources for single-ply membranes

For an excellent guide to single-ply membranes, contact the National Roofing Contractors Association (10255 W. Higgins Rd., Suite 600, Rosemont, Il. 60018; 708-299-9070). Their "Roofing & Waterproofing Manual" contains just about everything you'll need to know about roofing, including single-ply membranes. The manual costs $118 (non-members).

The Single-Ply Roofing Institute (104 Wilmont Rd., Suite 201, Deerfield, Il. 60015-5195; 708-940-8800) has a comprehensive list of various manufacturers, categorized by membrane type. There are too many

manufacturers to list in this article, but here's a selection of manufacturers whose products are generally available.   *—H. L.*

EPDM membranes
Firestone Building Products Co., 525 Congressional Blvd., Carmel, Ind. 46032; 800-428-4442

Resource Conservation Technology, Inc., 2633 N. Calvert St., Baltimore, Md. 21218; 301-366-1146

The Goodyear Tire & Rubber Co., Roofing Systems, 1144 E. Market St., Akron, Ohio 44316; 800-992-7663

Manville Corporation, Roofing Systems Div., P. O. Box 5108, Denver, Colo. 80217; 800-654-3103

CSPE membranes
JPS Elastomerics Corporation, P. O. Box 658, Northampton, Mass. 01061-0658; 413-586-8750

MBR membranes
U. S. Intec, Inc., P.O. Box 2845, Pt. Arthur, Tex. 77643; 800-624-6832

Owens-Corning Fiberglas Corp., Commercial Roofing, Fiberglas Tower, Toledo, Ohio 43659; 419-248-8000

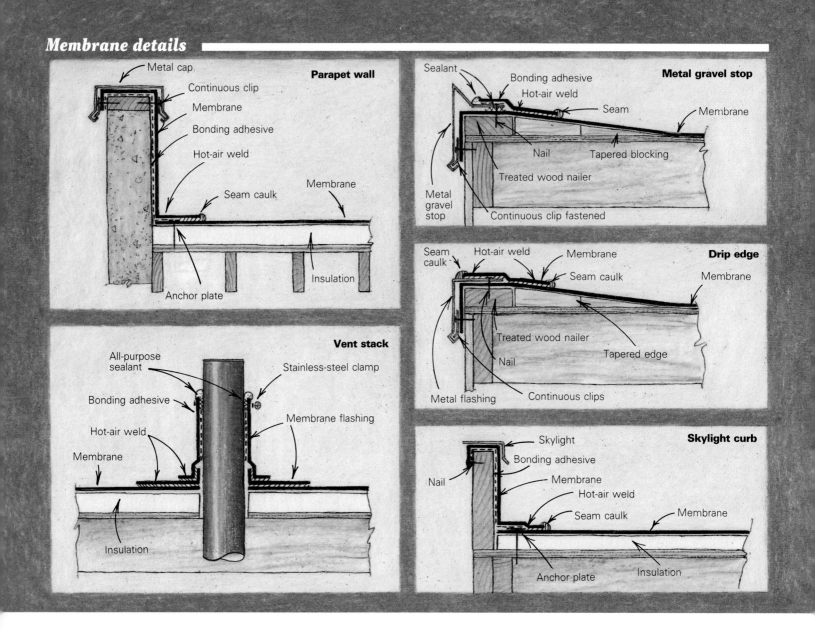

ventional built-up roofing, extreme care must be taken to ensure that no asphaltic products or debris are tracked onto the PVC membrane. A separator sheet should be used beneath the PVC membrane to ensure that it doesn't come in contact with the original membrane. Perimeter flashings must be made using special metal accessories precoated with PVC to which the roof sheet can be fused.

While the material comes in white, grey, and some colors, it tends to attract atmospheric dirt almost magnetically, so it can be difficult to keep clean where appearance is important. It might seem that the appearance of a roof membrane on a flat roof is less of a factor than the appearance of a gable-roof covering. But when an adjacent portion of the building overlooks the membrane, the view below shouldn't detract from the view beyond.

Modified bitumen membranes—As the quality of asphalt bitumen for roofing deteriorated, synthetic rubber and plastic membranes captured ever larger shares of the roofing market. The asphalt industry decided to fight back by researching ways to improve the performance of the asphalt. What they did was to modify raw asphalt with chemicals, resulting in a rubberized asphaltic material generally called "modified bitumen." The product is combined either with fiberglass or with non-woven polyester reinforcement to form a 160-mil roll product that's installed as a single layer, rather than being hot mopped in several succeeding layers as with conventional built-up roofing. With the reinforcement, the product is called MBR (modified bitumen, reinforced).

MBR membranes are extremely tough and puncture-resistant, so they're well suited to areas where there's the possibility of falling tree branches and other physical projectiles. At 1 lb./sf, they're about three times the weight of CSPE and EPDM, but are considerably lighter than a built-up roof (a 4-layer smooth surface built-up roof weighs about 1.5 lb./sf without gravel and 5 lb. to 6 lb./sf with gravel). Like other types of single-ply membranes, MBR membranes may be applied to roofs of any slope, from dead flat to absolutely vertical. An MBR membrane generally costs slightly less than rubber or PVC membranes.

Field seams in MBR roofs are made with the same techniques used to install the membrane itself: heat welding, hot mopping or cold adhesives. The basic color of MBR sheets is black, but some manufacturers offer membranes with a colored granular surface. Loose granules in a matching color can be sprinkled onto the field seams to achieve a uniform color or over the entire roof surface.

Because the asphalt is modified, MBR sheets are more flexible than sheets of conventional roll roofing and can be formed around corners without tearing, particularly when they're softened with a torch. But MBR sheets do have limits, and can be difficult to form into the complex shapes that are needed to flash some penetrations. Perimeter and penetration flashings are made using the same membrane used for the roof. Like conventional built-up roofs, MBRs are susceptible to concentrated acids and to hydrocarbons.

Unfortunately, some MBR roofing requires the use of installation techniques reminiscent of the old hot-kettle methods used for built-up roofs, methods that other single-ply membranes have done away with. Several different types of modified bitumens are available today, but two types predominate: APP (axactic polypropylene) and SBS (styrene butadiene styrene). The difference between them is largely the method of installation.

APP modified bitumen membranes are "heat applied." A roll of membrane material is posi-

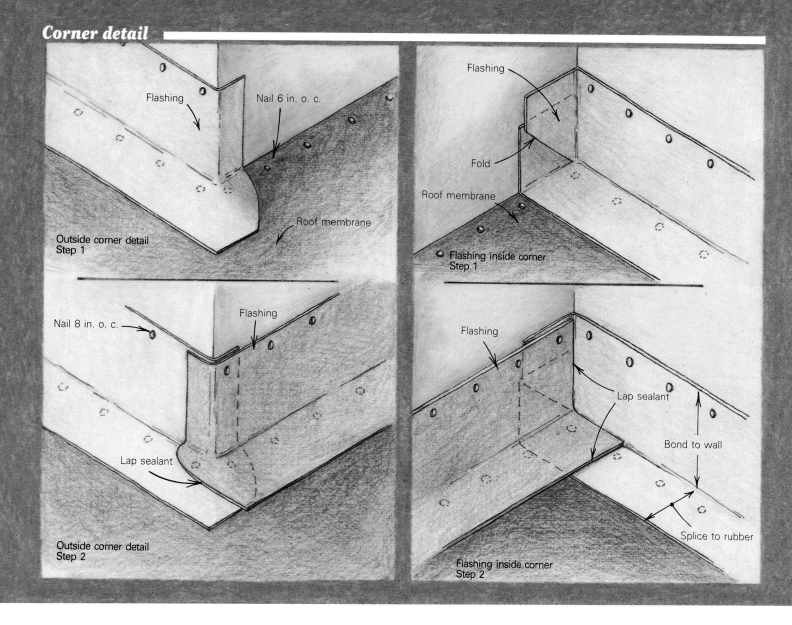

Flashing — Nail 6 in. o. c.

Outside corner detail
Step 1

Roof membrane

Flashing

Fold

Roof membrane

Flashing inside corner
Step 1

Nail 8 in. o. c. — Flashing

Lap sealant

Outside corner detail
Step 2

Flashing

Lap sealant

Bond to wall

Splice to rubber

Flashing inside corner
Step 2

tioned so that it will unroll into its final position on the roof. The underside of the roll is then heated with an electric heating element or a gas torch until the material softens and begins to flow. Then the membrane is rolled out, and the melted underside becomes the adhesive (photo, p. 103). For small roofs the torch may be hand-held, but for large roofs there are special carts to hold the roll and a built-in row of torch nozzles so that a uniform flame is applied across the entire width of the roll. The equipment and the process are proven, and they're safe in the hands of skilled and careful installers. But the process requires high temperatures, so torch-applied MBR roofing is usually not recommended over wood decks or combustible insulation materials.

An SBS membrane may be applied using a flood coat of conventional hot asphalt as an adhesive, or by using a proprietary cold adhesive. The hot-mopped installation brings back the asphalt kettle and open flame associated with built-up roofing, and so may not be attractive to people who were glad to bid adieu to such equipment. In addition to being smelly and dangerous, hot asphalt is just plain messy. It's tough to use without spilling some of the black goo where it is least wanted.

Torch-applied APP, even though subject to some possibility of fire as a result of the torches, is neater than hot-mopped SBS. Adhesive applied SBS is equally neat, and does not require any open flames on the job site.

Application is for the pros—While almost anyone can buy a plug of asphalt and some rolls of roofing felt in order to do a built-up roof, most of the single-ply products are available only to franchised applicators. While this means that a builder generally can't buy the material and do it himself (an approach I don't recommend, because it precludes any warranty from the manufacturer), it also means that when you find a licensed applicator, the full resources of the manufacturer are also available for technical advice.

If you insist on working with the material yourself, and some builders do, you won't find it at the local lumberyard. Instead, check with a roofing wholesaler, or go to a franchised applicator and purchase surplus sheeting. Remember, though, that with CSPE and EPDM membranes, you'll also need the compatible adhesives and accessories. The drawings above show samples of membrane details commonly encountered in residential work.

Patching a hole in an EPDM membrane is just about like patching an inner tube. After you find the leak, cut a patch, clean the surrounding area and fasten it down with adhesives. CSPE and PVC membranes have to be chemically treated before a patch can be applied, however, because the membrane itself has to be softened up in order for the patch to be heat-fused on. Any serious repairs of single-ply membranes are best left to those who put the system in.

One design note. Any kind of single-ply membrane makes an excellent vapor barrier, so make sure there's plenty of air circulation between any insulation and the membrane to prevent moisture from being trapped in the roof assembly. Alternately, you can keep moisture from the roof cavities by providing a vapor barrier beneath the interior ceiling surfaces. Protected membranes, however, are one exception. Because this puts the membrane on the warm side of the insulation, there's no need for a separate vapor barrier. □

Harwood Loomis is a consulting architect specializing in building envelopes, roofing and waterproofing. Photos by the author except where noted.

Decorative Roof Shingles

Asphalt-shingle roof by Lance Linfoot, Laytonville, Calif.

Photo: Jim Nostdal

Cedar-shingle roof by Robin Howard, Dayton, Wash.

Photo: David S. Palmer

Slate-shingle roof of the Harvey Firestone farm, moved from its original site in Ohio to Greenfield Village in Dearborn, Mich.

Slate roof from recycled slates by Tom Duca, Essex, N. Y.

Studio and sculptural-shingle roof by Don Beard, Ft. Collins, Colo.

Trulli Amazing

The story goes that when these houses, known as *trulli*, were built in the southeastern Italian region of Puglia, their limestone roofs were constructed without mortar because the feudal lord could tax only buildings with permanent roofs. Before the tax collector made his rounds, the roofs would be disassembled, and no tax could be levied.

The roofs are held in place by gravity and the lateral opposition of the limestone. Italian law protects the *trulli* as national monuments. Photos by John Ferro Sims.

INDEX

The articles in this book originally appeared in *Fine Homebuilding* magazine. The date of first publication, issue number and page numbers for each article are given at right.